Volume Two

ENERGY AND ENVIRONMENTAL CHEMISTRY

Acid Rain

Edited by
LAWRENCE H. KEITH

ANN ARBOR SCIENCE
THE BUTTERWORTH GROUP

jwl 9-26-83

Energy and Environmental Chemistry—they are inextricably entwined. As we progress toward an increasingly industrialized and energy hungry world the demands of producing more and more energy are going to require greater care if we are to maintain a clean environment. There is no question that we will continue to produce more energy—progress demands it and people demand progress. The only question is how much we will let our continuous quest for energy affect our environment.

Learning about the effects of energy production—directly or indirectly—on the environment is one of the first steps to controlling adverse effects of that production. That's what these volumes are all about. They do not cover it all—do not even come close; the subject area is much too large and complex and we are only in the early stages of learning about the interrelations of energy and environmental chemistry. But it is hoped that the information in these books will bring us one step closer to understanding some of these relationships and hence to ultimately controlling unwanted pollution from energy production.

Lawrence H. Keith

ACKNOWLEDGMENTS

This is the second of a series of volumes entitled, "Energy and Environmental Chemistry." These volumes are collected papers from distinguished authors on the North American continent and abroad, and span several national and international meetings beginning with the Second Chemical Congress of the North American Continent in Las Vegas, Nevada (1980).

These symposia were jointly sponsored by the American Chemical Society's Committee on Environmental Improvement and the Division of Environmental Chemistry. However, it was only with the help of John I. Teasley and Gary E. Glass from the U.S. Environmental Protection Agency laboratory in Duluth, Minnesota, that various sessions of these symposia were organized. Without their help and commitment, the symposia, and hence this volume, could never have been accomplished.

Finally, I wish to express my appreciation to the many authors who, by their hard work, dedication and commitment, provided the main substance of this work—scientific facts and evaluations on a growing national and international problem: acid rain.

Lawrence H. Keith's current technical interests continue to center around analyses of organic pollutants in the environment, with emphasis on developing new methods or improving on old ones. Techniques for the safe handling of carcinogenic and/or extremely toxic materials are also an important aspect of Dr. Keith's current research efforts.

Dr. Keith was formerly involved with the selection of many of the U.S. Environmental Protection Agency's initial 129 Priority Pollutants, and he also helped to formulate some of the initial methodology for analyzing for these pollutants. He is presently involved in the selection of representative compounds and methodologies for the Appendix C Priority Pollutants and for the synfuel industry.

A member of the American Chemical Society's Division of Environmental Chemistry Executive Committee, Dr. Keith has served as secretary, alternate councilor, program chairman and chairman of the division. He is also a past chairman of the Central Texas Section of the American Chemical Society and past secretary and councilor of the Northeast Georgia Section of the American Chemical Society.

In other professional activities, Dr. Keith served as Vice-Chairman of the Gordon Research Conference on Environmental Sciences: Water, and is currently a delegate of the U.S. National Committee to the International Association of Water Pollution Research. He is also a member of the National Research Council Committee on Military Environmental Research. Dr. Keith edited the two-volume *Advances In the Identification & Analysis of Organic Pollutants In Water*, published by Ann Arbor Science.

He and his wife, Virginia, reside in Austin, Texas.

Dedicated to the Memory of

Steven Douglas Keith

He knew the happiness of youth,
the love of a family,
the freedom of the birds
and the agony of hell.
He died doing what he
loved best—flying.

CONTENTS
VOLUME 1

xi

CONTENTS
VOLUME 2

Part 1: Point Source Effects

Part 2: Regional Effects

PART 1

POINT SOURCE EFFECTS

CHAPTER 1

THE EFFECTS ON THE ENVIRONMENT OF FOSSIL FUEL USE BY ELECTRIC UTILITIES

E. Kleber, A. Dasti and A. Gakner
Federal Energy Regulatory Commission
Washington, DC 20426

Recent technical reports have highlighted growing concern about the "acid-rain" problem. Brezonik et al. [1] made pH determinations on rain-water in Florida during a recent year, where isolated measurements ranged to a low of 3.76, and volume-weighted measurements ranged from an average of 4.60 to a high of 5.80. Lewis and Grant [2] made measurements in Boulder County, Colorado, over a recent three-year period which indicated a drop in pH from 5.4 to 4.6 over that time span.

The causes and implications of these measurements have been argued vigorously, and much attention has been directed to the electric utility industry, especially because of its combustion of coal with a high sulfur content for power generation. However, the inadequacy of data sufficient for reaching scientific conclusions has become apparent, and led to the recent enactment of federal legislation directed toward more thorough evaluation of cause and effect. This legislation is contained in the Energy Security Act [3], signed June 30, 1980. Title VII of the act is the Acid Precipitation Program and Carbon Dioxide Study.*

---

*The act established a ten-year program to carry out the provisions of Title VIIA, with the formation of a Task Force, having as joint chairmen the Secretary of Agriculture, the Administrator of the Environmental Protection Agency, and the Administrator of the National Oceanic and Atmospheric Administration. The latter is also named as the director of the research program established by this act.

Subtitle A of Title VII of the act is known as the Acid Precipitation Act of 1980. One objective is the development of a comprehensive research plan to identify causes and effects, and to identify actions to limit or ameliorate the harmful effects of acid precipitation.*

Acid precipitation may be significantly affected by the electric utilities' use of sulfur-bearing fossil fuels. The data that follow are based on information available in Department of Energy (DOE) records.

In 1972 the Federal Power Commission (now the Federal Energy Regulatory Commission) initiated a new data-collection system, which required filing of monthly reports (form 423) to the DOE of cost and quality of fuels for electricity-generating plants. In its present form, this report requires the filing of information on fossil-fuel shipments received by each electric-power producer (both private and public) for each of its electric-generating plants with a total combined (steam-electric, combustion turbine and internal combustion engine) generating capacity of 25 MW or more. The information provided by the utility for each delivery of fuel during the reporting month includes the following:

Type of plant: steam turbine, combustion turbine, internal combustion engine
Purchase type: (coal and oil) spot purchase, new contract, contract under which
    price is changed because of automatic price adjustment, all other contracts
Contract expiration: within the next 24 months
Fuel type: coal, oil, gas
Coal mines: type (underground or surface);
        origin (state and Bureau of Mines coal-producing district)
Fuel source: coal—name of mine and county of origin;
        oil—supplier and refinery or port of entry
        gas—pipeline or distributor, producing area or port of entry
Quantity received: tons, barrels, MCF
Quality received: Btu content, sulfur content, ash content
Delivered price: cents per million Btu.

The data as received are entered on a computer which makes possible retrieval in diverse ways. Reports summarizing some of the data are issued monthly (e.g., [4]). Annual reports, which include a variety of special studies, also are published. The first annual report was published in 1975 [5] covering the years 1973 and 1974. Subsequent reports have been expanded to include

---

*Section 704 provides for the development of the plan and its submission in draft form to the Congress and for public review by December 30, 1980, and in final form by April 15, 1981. The plan is to include programs in 14 subareas of research, development and assessment.

additional, most frequently, requested analyses, and have been published annually since then [6].

With this wealth of data available, attention has been focused on the sulfur content of the fuels, and some of the possible implications of combustion of these fuels with the corresponding introduction of sulfur-containing species into the atmosphere. These data are shown here in a series of tables which represent the maximum potential for sulfur dioxide emissions into the atmosphere by fuel combustion. These are based on the assumption that the entire sulfur content of the fuel is converted into sulfur dioxide, all of which finds its way into the atmosphere. Ignored for the purposes of this calculation are estimations of any quantities that do not enter the atmosphere by virtue of either retention of any portion of the sulfur in the ash, or the removal of sulfur oxides by stack gas scrubbing.

The data in Table 1 put into perspective the reliance of the electric utility industry on coal as a source of fossil fuel for the generation of electricity. The data cover the period since 1962 in five-year increments. Generation from fossil fuel sources was based on the Btu content of the fuels; nuclear and hydro generation were calculated from kWh statistics, using a conversion factor of 10,000 Btu/kWh. The major portion of the increase in electricity generation in the decade of the 1960s was from the increased use of fossil fuel; in the 1970s, following the oil embargo, increases in total electricity generation and in fossil-fuel demand have been very small.

Detailed data for the period 1975-1979 are also shown in Table 1. Whereas total generation increased at a 3.9% annual rate, coal-fired generation increased at a 6.2% annual rate. (The data included for 1980 are based on extrapolations from five months of statistics.) The decrease in 1978 in coal-fired generation does not represent a real trend—it was caused by an unusually long (four-month) strike by coal miners in the East. The drop in oil-fired electricity generation is real, caused by the deliberate push for economic and political reasons to switch away from oil as a fuel source. The dip in hydro-generation in 1977 was caused by a severe water shortage in the western United States that year.

Total national emissions of sulfur dioxide, using the aforementioned assumptions, are shown in Table 2. In general, one can observe that the increasing tendency to use coals of lower sulfur content has resulted in relatively constant levels of total sulfur dioxide emissions, despite significant increases in coal quantities delivered. Contributions to total sulfur dioxide emissions by oil utilization have been much smaller, and on a national basis, do not alter the totals markedly.

For comparative purposes, it may be noted that the U.S. Environmental Protection Agency (EPA) has estimated [7] that the total emissions of SO_x (x not specified) including those from nonelectric sources have ranged from 24,500 x 10^3 tons in 1950 to 29,900 x 10^3 tons in 1976, with a high of

Table 1. Generation (10^{15} Btu) by the Electric Utility Industry: By Primary Sources of Energy

	Coal	Oil	Gas	Total Fossil	Nuclear	Hydro	Grand Total[a]
1962	4.66	0.58	2.03	7.27	0.02	1.75	9.04
1967	6.90	1.01	2.83	10.75	0.08	2.30	13.13
1972	8.39	3.14	4.11	15.63	0.58	2.81	19.04
1977	10.55	3.83	3.19	17.57	2.51	2.20	22.28
1975	9.33	3.07	3.11	15.51	1.73	3.00	20.24
1976	9.86	3.31	3.03	16.20	1.91	2.84	20.95
1977	10.55	3.83	3.19	17.57	2.51	2.20	22.28
1978	10.13	3.72	3.23	17.08	2.76	2.81	22.65
1979	11.89	3.02	3.35	18.27	2.55	2.80	23.63

[a]Includes all other sources.

Table 2. Sulfur Dioxide Emissions

	Coal			Oil			
	10^6 ton	% S	Total SO$_2$ (10^3 ton)	10^6 bbl	% S	Total SO$_2$ (10^3 ton)	Total SO$_2$ (10^3 ton)
1975	431.1	2.2	19,000	468.9	1.0	1,600	20,600
1976	454.9	2.07	18,800	508.3	1.0	1,800	20,600
1977	490.4	1.95	19,100	584.4	0.9	1,800	20,900
1978	476.2	1.79	17,000	568.6	1.0	2,000	19,000
1979	556.6	1.73	19,200	489.0	1.0	1,700	20,900
1980[a]	630	1.58	19,900	385	1.0	1,300	21,200

[a]Extrapolated from data for the first five months of the year.

32,700 x 10^3 tons in 1973. The EPA attributes the principal source of the SO_x to fuel combustion from stationary sources.

National data can be disaggregated to provide regional information. Summaries of total sulfur dioxide emissions (again with assumptions utilized) as compiled by geographic regions are shown in Table 3. Total contributions for each region in 1979, as a percentage of the national totals are shown in Table 4. Aggregations of the data may be made in many ways: one such would show that power plants east of the Mississippi were responsible for 83% of total sulfur dioxide emissions nationally while providing only 62% of electricity generation.

One feature of the regional data is the indication of the relative constancy of total sulfur dioxide emissions from electric utility power plants over the past five years. The only exceptions are in three geographic regions with less than 10% of total emissions—New England, West South Central, Mountain— where there has been some increase in recent years. Further detailed data analyses may be made using any mode of aggregation to provide an input into specific studies such as those cited [1, 2].

It is interesting to combine sulfur dioxide emission data with rainfall data. If one assumes that: (1) all the sulfur dioxide emitted to the atmosphere is quantitatively converted to sulfur trioxide; (2) the latter is quantitatively converted to sulfuric acid by contact with rainwater; and (3) all the sulfuric acid reaches the earth as acid rain with no precipitation in dry form as neutral sulfate dusts; then the maximum acidity of the corresponding rain due to sulfuric acid can be calculated. Table 5 shows the results of these calculations using 1977 data for rainfall. West of the Mississippi, the calculations of sulfuric acid content in rainfall yield concentrations of 0.5 to 2.5 x 10^{-6} molal; east of the Mississippi, the concentrations are calculated to range from 2 to 17 x 10^{-5} molal. Calculations may be made for any geographic area: for example, data for the state of Florida, using all of the assumptions above, show a sulfuric acid concentration of 5 x 10^{-5} molal.

Future trends may also be forecast with the aid of these data.

1. Coal usage will accelerate significantly. A recent analysis prepared in this office [9] indicates that demand due to new plants will be for an additional 170,000 tons of coal in 1983 (a 25% increase) and 395,000 tons by 1988 (>50% increase). (These do not take into account any reduction in demand due to retirement of old plants.)
2. There will be an increasing trend toward use of low-sulfur coal, to the extent that much more low-sulfur Western coal can be made available. The net effect of these two factors will be a relative constancy in total emissions of sulfur dioxide to the atmosphere, paralleling the trends of recent years.
3. The effects of the installation of flue gas desulfurization systems may be ignored for the near-term future. Data summarized by the EPA [10] indicate that, as of March 1980, 10% of the capacity of all coal-fired units have the capability to scrub sulfur dioxide. Operationally, however, the percentage of coal-fired scrubbed electric generation is much less.

Table 3. Sulfur Dioxide Emissions, Regional (10^3 ton)

	New England			Middle Atlantic			East North Central		
	coal	oil	total	coal	oil	total	coal	oil	total
1975	60	170	230	2430	380	2810	7810	65	7875
1976	35	230	265	1910	390	2300	7630	60	7690
1977	45	225	270	1950	420	2370	7500	80	7580
1978	35	345	380	1670	435	2105	6680	90	6770
1979	55	360	415	2010	360	2370	7560	70	7630
1980[a]	85	320	405	2120	290	2410	7340	70	7410

	West North Central			South Atlantic			East South Central		
	coal	oil	total	coal	oil	total	coal	oil	total
1975	2000	30	2030	2700	670	3370	4110	60	4170
1976	2200	30	2230	2700	690	3390	3690	130	3820
1977	2230	30	2260	2990	730	3720	3590	140	3730
1978	2000	30	2030	2690	720	3410	3050	140	3190
1979	2180	20	2200	2920	620	3540	3320	90	3410
1980[a]	2340	5	2345	3160	500	3660	3860	40	3900

	West South Central			Mountain			Pacific		
	coal	oil	total	coal	oil	total	coal	oil	total
1975	110	40	150	330	25	355	40	140	180
1976	160	80	240	430	15	445	40	100	140
1977	240	130	370	520	25	545	55	180	235
1978	320	140	460	490	20	510	95	90	185
1979	490	80	570	600	20	620	85	100	185
1980[a]	650	30	680	800	10	810	75	90	165

[a]Extrapolated from data for the first five months of the year.

Table 4. Regional Contributions to Sulfur Dioxide Emissions

Region	SO_2 Emissions (% of total)	Sulfur Content of Fuel (%)	
		Coal	Oil
New England	2.0	2.56	1.4
Middle Atlantic	11.3	2.03	1.0
East North Central	36.4	2.30	0.6
West North Central	10.5	1.49	0.9
South Atlantic	16.9	1.55	1.4
East South Central	16.3	2.29	2.3
West South Central	2.7	0.59	0.9
Mountain	3.0	0.54	0.7
Pacific	0.9	0.87	0.3

Table 5. Conversion of Sulfur Dioxide Emissions to Acid Rain [8]

Region	Area $(10^3 \ mi^2)$	1977 Rainfall (in.)[a]	Calculated H_2SO_4 Concentrations $(10^{-6} \ molal)$
New England	66.6	48.61	1.79
Middle Atlantic	102.7	48.29	10.2
East North Central	248.3	38.26	17.1
West North Central	517.4	31.76	2.64
South Atlantic	278.9	45.66	5.75
East South Central	181.9	55.81	7.87
West South Central	438.8	31.06	0.58
Mountain	863.8	12.76	1.05
Pacific	323.9	24.17	0.64

[a]Areally weighted.

The conclusions from these data are that utility-caused sulfur dioxide emissions have been relatively constant in recent years, will continue to be so in the near future, and that increasing acidity in rainwater cannot be attributed directly to fossil fuel utilization by the electric utilities.

ACKNOWLEDGMENT

The opinions expressed are strictly those of the authors, and do not in any way reflect the views of the Federal Energy Regulatory Commission.

REFERENCES

1. Brezonik, P. L., E. S. Edgerton and C. D. Hendry. "Acid Precipitation and Sulfate Deposition in Florida," *Science* 208:1027 (1980).
2. Lewis, Jr., W. M., and M. C. Grant. "Acid Precipitation in the Western United States," *Science* 207:176 (1980).
3. PL 96-294, the Energy Security Act.
4. "Monthly Report of Cost and Quality of Fuels for Electric Utility Plants—April 1980," DOE/EIA-0075 (80/04).
5. "Annual Summary of Cost and Quality of Steam Electric Plant Fuels, 1973 and 1974," Federal Power Commission, Bureau of Power (May 1975).
6. "Cost and Quality of Fuels for Electric Utility Plants: Energy Data Reports," DOE/EIA-0191.
7. "National Air Pollutant Emission Estimates, 1940-1976," U.S. EPA-450/1-78-003 (July 1978).
8. "State, Regional, and National Monthly and Annual Total Precipitation," National Oceanic and Atmospheric Administration (August 1978).
9. "Status of Coal Supply Contracts for New Electric Generating Plants," 1979 supplement (to be published, Energy & Fuels Analysis Branch, FERC).
10. "Utility FGD Survey, Jan-March 1980," U.S. EPA-600/7-80-029b (May 1980).

CHAPTER 2

PLUME WASHOUT AROUND A MAJOR COAL-FIRED POWER PLANT: RESULTS OF A SINGLE STORM EVENT

N. C. J. Chen and R. E. Saylor

Oak Ridge National Laboratory
Engineering Technology Division
Oak Ridge, Tennessee

S. E. Lindberg

Oak Ridge National Laboratory
Environmental Sciences Division
Oak Ridge, Tennessee

INTRODUCTION

As part of a larger study on the Meteorological Effects of Thermal Energy Releases (METER) [1], field experiments have been conducted since 1978 using a dense raingauge network (49 sites, 1500 km^2) to evaluate the potential precipitation modification by the heat and moisture releases from the Bowen Electric Generating Plant and to evaluate the impact of emission on precipitation chemistry [1, 2]. The power plant, located ~ 64 km northwest of Atlanta, Georgia, has a total electric generating capacity of 3160 MW by the daily combustion of ~ 35,000 tons of coal, discharging heat and combustion emissions through two 305-m precipitator-equipped smokestacks and four 119-m natural-draft cooling towers. The plant is situated in an ideal study location (gentle rolling terrain without multiple sources) and is representative of near-future coal combustion technology.

11

Reported here are some results of a plume washout field experiment performed during the first three weeks of December 1979 in the vicinity of the power plant (termed the WISPE experiment, Winter Study of Power Plant Effects). Precipitation was collected using a network of HASL-type collectors for analysis of plume washout effects on rain chemistry. The Pennsylvania State University (PSU) used an instrumented aircraft to measure in-plume sulfur dioxide concentration, sulfur aerosol particle size and number density, and various meteorological parameters. Battelle Pacific Northwest Laboratory (BPNL) provided necessary wind profiles from data gathered from fixed meteorological towers and pilot balloons.

This chapter describes the analysis of the local plume washout effect on rain chemistry (major and trace constituents) from analysis of wetfall-only precipitation samples and onsite meteorological data taken during a single 0.67-cm rain event on December 13, 1979. A number of papers [3-9] have indicated the importance of sulfate plume washout from power plant emissions. The WISPE data reported here support the existence of local plume washout of sulfate.

EXPERIMENTAL

The wetfall chemistry network was a subset of the existing METER-ORNL raingauge network (Figure 1) consisting of a matrix of 19 automatic HASL-type collectors placed within a 300-km^2 area included in the larger network (~ 1500 km^2) of a 7 x 7 matrix of 49 automatic raingauges [10,11]. Throughout the experiment on December 13, the power plant electric output was ~ 1300 MW, coal combustion $\sim 13,000$ tons with $\sim 1.5\%$ sulfur content. Rainfall was monitored continuously at each raingauge site. Similarly, wind speed and direction were monitored continuously by windsets and routinely by pilot balloons and visual observations.

A polyethylene bottle-funnel combination was placed in each HASL collector and remained sealed against dry deposition for two days prior to the onset of the storm. (The design, operation and efficiency of a similar sampler have been described in detail by Lindberg et al. [12].) The samples were retrieved as rapidly as possible (< 7 hr) after the cessation of rain and stored at 4°C prior to analysis. A combination reference-pH electrode was used to measure pH within 24 hr of sample collection. Major constituent analyses were performed within 3 days using a Technicon autoanalyzer and a modified methylthymol blue colorimetric method for sulfate [13], a cadmium reduction method for nitrate, and an indophenol method for ammonium ion (Technicon Industrial Methodologies No. 158-71W and 154-71W, Technicon, Inc., Tarrytown, NY). Trace metals were analyzed on

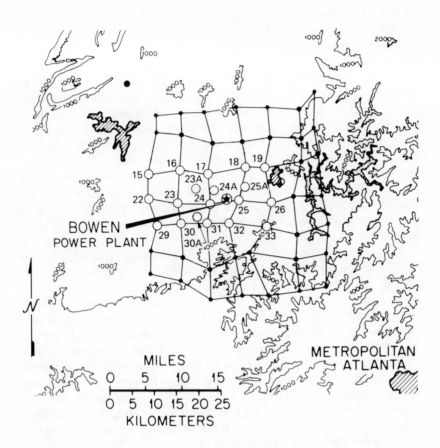

Figure 1. Map showing the chemistry network (open circles with the site number indicated) in relation to the raingauge network (solid dots). The topography of northwest Georgia (with the location of Bowen Power Plant and several other major features) is superimposed with the values of the contour lines indicating feet above mean sea level.

separate aliquots acidified on collection to 0.1 N Ultrex HNO_3 using graphite furnace atomic absorption spectrophotometry. Details on sulfate and trace metal analytical methods including discussions of precision and accuracy and the interpretation of acid leachable trace metal concentrations in rain are presented elsewhere [9, 12].

RESULTS

Meteorological Parameters

A comparison of rainfall statistics (Table 1) between the full METER raingauge network and the subset of this network which included the rain chemistry samplers indicated no significant differences in the means or standard deviations, suggesting that the rainfall amounts in the chemistry network for the December 13 storm were comparable to those of the larger region. Within the rain chemistry network, however, there were well defined trends in both intensity (Figure 2) and rainfall volume (not shown) of increasing values from NW to SE across the network. This variation in rain volume complicates analysis of the plume washout effect because of the inverse relationship between volume and rain concentration [9, 14, 15] as discussed below.

To determine plume location during the storm, high-frequency fluctuations of winds, measured at the meteorological tower (~150 m above the ground), were smoothed by a seven-point weighting scheme [16] at 10-min intervals and the derived plume direction was assumed to spread by $10°$ on each side of the plume centerline. Each 10-min time interval that the plume remained over a sampling station was summed to obtain a total plume residence time during rain. The residence times and event durations for those sites in the target area are listed in Table 2. This analysis established the target area to be the eastern half of the chemistry network. The predominant downwind direction was identified to be the SE sector of the network, over which the plume was situated for ~40% of the storm duration. Visual observation of the plume confirmed that the wind data measured at the meteorological tower closely defined its location.

In an analysis of power plant effects on rain chemistry, variations in rain concentration or wet deposition rates between target and control areas may be ascribed to both plume scavenging and rainfall volume effects; thus, the "ideal" storm should produce uniform rainfall amounts at all sites. The variation in rainfall between the eastern (target) and western (control) areas in the

Table 1. Characteristics of December 13 Storm[a]

	METER-ORNL Raingauge Network (7 x 7)	Chemical Network (subset METER-ORNL)
Mean Rainfall (mm)	6.74 (49 sites)	6.76 (17 sites)
Standard Deviation (mm)	1.52	1.43
Coefficient of Variation	23%	21%

[a]Type: cold front; rain period started at 0830, ended at 1800.

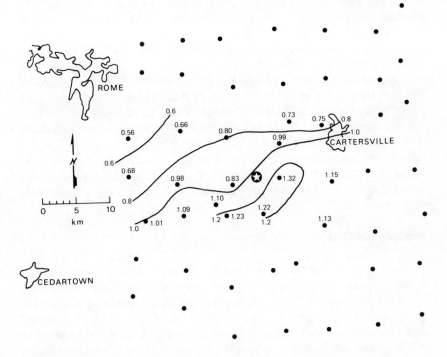

Figure 2. Rainfall rate contours in mm/hr for the storm of December 13, 1979, over the chemistry network.

Table 2. Plume Residence Time During Rain

Site	Plume Residence Time (hr)	Event Duration
18	0	1030–1730
MET	0.17	1030–1730
19	0.17	1000–1730
25		Inoperable
26	1.83	1000–1700
32	2.00	1030–1700
33	3.00	1030–1810

network was such that the mean rainfall rate in the target area exceeded that in the control area by ∼10% (Table 3). This results in some uncertainty in defining cause and effect for any differences in wet deposition between target and control areas. If the presence of the power plant influences rainfall volume (this is the hypothesis of the original METER study [1]) as well as chemistry in the target area, then any differences in rainfall concentrations or wet deposition between target and control areas may be ascribed to the emission source. The METER rainfall study is addressing the potential power plant induced modification with extensive statistical analyses of rainfall on a storm event basis [17]. Any significant increase in rain concentrations or wet deposition in the target area cannot be defined simply as scavenging of plume constituents by precipitation. With this caveat in hand, we will describe the wetfall chemistry in the target and control areas and offer one possible hypothesis to explain the significant difference in the sulfate deposition rates.

Precipitation Chemistry

There were no statistically significant differences in precipitation concentration of major or trace components between the target and control areas from the December 13 storm (Table 3). Only sulfate exhibited a difference in the mean concentrations of $\geq 15\%$. The concentrations of both the major and trace constituents in target area rain did not reflect a major influence of the emission source when compared with published data. Concentrations were generally representative of values reported for rural areas [9, 18–20].

Despite a higher rainfall rate in the target area relative to the control, only SO_4^{2-} and NO_3^- exhibited significantly ($p > 0.05$) higher wet deposition rates in the target area than in the control area. Mean target area deposition rates for both constituents were roughly 30% above those of the control area; the majority of the remaining constituents exhibited mean target area deposition rates within ∼15% of control values. The target area mean wet deposition rates of Ni and H^+ exceeded those in the control area by 20–24%, but did not reflect a significant difference in the t statistic.

DISCUSSION

Deposition rates are the product of rainfall intensity and concentration. The significant differences in deposition rates of SO_4^{2-} and NO_3^- to target and control areas can, thus, be accounted for by differences in either or both components. There were no significant differences in the rainfall intensities or in concentrations of SO_4^{2-} and NO_3^-, although the target area values ex-

Table 3. Statistical Summary of Precipitation Concentrations and Wet Deposition Rates for Target and Control Areas for December 13 Storm

	Rainfall Amount (mm)	Rainfall Amount (mm/hr)	Parameter											
			H⁺	SO₄²⁻	NO₃⁻	NH₄⁺	Al	Fe	Mn	Cu	Pb	Zn	Ni	Cr
			(μmol/L)						*(nmol/L)*					
			Concentrations											
$\bar{X}_T$[a]	7.03	0.99	31.65	16.61	8.35	10.93	0.59	0.17	9.31	9.45	20.45	80.58	20.06	2.08
SE_T[b]	0.58	0.09	4.36	1.10	0.41	1.41	0.06	0.03	0.94	1.84	1.48	22.05	13.01	0.19
$\bar{X}_T/\bar{X}_C$[c]	1.10	1.11	1.10	1.15	1.09	1.03	0.92	0.97	0.97	0.90	0.98	0.93	1.13	1.06
t[d]	0.86	0.88	0.46	1.30	0.73	0.16	0.39	0.11	0.16	0.30	0.09	0.26	0.20	0.38

	H⁺	SO₄²⁻	NO₃⁻	NH₄⁺	Al	Fe	Mn	Cu	Pb	Zn	Ni	Cr
	(nmol/cm²-hr)						*(pmol/cm²-hr)*					
	Wet Deposition Rates											
$\bar{X}_T$	3.00	1.60	0.81	1.03	0.06	0.02	0.89	0.90	1.99	8.20	2.06	0.21
SE_T	0.28	0.05	0.04	0.06	0.01	0.003	0.05	0.14	0.16	2.54	1.48	0.03
$\bar{X}_T/\bar{X}_C$	1.24	1.29	1.26	1.13	1.01	1.06	1.09	0.92	1.15	1.09	1.20	1.16
t	1.30	5.11[e]	3.40[e]	1.01	0.04	0.16	0.57	0.21	1.25	0.24	0.26	0.69

[a] Mean value in target area.
[b] Standard error of the mean in target area.
[c] Ratio of mean value in target to mean value in control areas.
[d] Statistic calculated from Student's test of the significance of the difference between the mean values in the target and control areas (unequal sampling size, 14 degrees of freedom).
[e] Probability <0.05.

ceeded control values by ~ 10% (intensity and NO_3^-) to 15% (SO_4^{2-}). This suggests that intensity and concentration differences contribute roughly equally to the excess target area deposition rate of NO_3^-, while concentration differences were somewhat more important to the excess target area deposition rate of SO_4^{2-}.

The areal deposition pattern of SO_4^{2-} and the in-plume sulfur concentration data collected by PSU during this experiment provided some evidence for precipitation scavenging of SO_4^{2-} from the combustion plume. The deposition rates of sulfate (Figure 3) and nitrate (not shown) demonstrated well-defined spatial patterns suggesting possible plume washout effects, whereas the deposition rates of the trace metals and hydrogen and ammonium ions occurred in random patterns (not shown). In general, deposition of sulfate produced a pattern fanning out from the bottom row of the network with increasing deposition rates from west to east. The gradient was weaker in the control (western) area and stronger in the target (eastern) area, with the maximum deposition rate occurring at site 33, located southeast of the power plant.

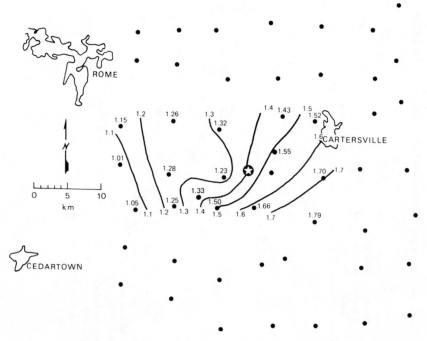

Figure 3. Contours of sulfate ion deposition rate in nmol/cm²-hr for the storm of December 13, 1979.

The extent to which the plume could have contributed to the excess deposition of sulfate can be estimated using a simplified washout model. Excess deposition rate was defined as the difference in sulfate deposition rate between each site in the target area and the mean deposition rate of the control area (1.23 ± 0.14 nmol/cm^2-hr). The model may be most accurately applied to target area site 33 (Figure 3) because the in-plume sulfate aerosol data were collected at approximately the same downwind distance from the stack as this site (~ 12 km). In addition, this site exhibited the highest sulfate deposition rate (Figure 3) and longest plume residence time (Table 2), and yet not the highest rainfall rate, making it a good candidate for the maximum possible plume influence. The parameters used for calculation are: a 500-μm mean raindrop radius, a collection efficiency of ~ 0.1 [21], a plume thickness of ~ 300 m at 12 km downwind of the stack, and rain intensity of ~ 1 mm/hr. Sulfate aerosol exhibited a 1-μm mean radius in the plume, a density of ~ 2 g/cm^3, and a particle number density of ~ 1/cm^3, as measured during the WISPE experiment by the PSU airplane 12 km downwind of the power plant.

The SO_4^{2-} aerosol washout was estimated to be ~ 0.4 nmol/cm^2-hr based on this model, which accounts for $\sim 70\%$ of the observed excess deposition rate at site 33 (0.56 nmol/cm^2-hr). This value is approximate because of the assumptions involved, but does indicate the potential contribution of the plume to sulfate deposition. The washout effect is probably underestimated in the calculations because scavenging of SO_2 is not taken into account. On the other hand, the higher rainfall rate at this site (1.13 mm/hr) compared to the control area (mean 0.87 mm/hr) also accounts for some fraction of the excess in the sulfate deposition rate.

The possible influence of the plume can be further illustrated by plotting the relative excess deposition rate of sulfate at each site versus the plume residence time over each site (Figure 4). The relative excess deposition rate was calculated by normalizing the target area excess deposition rates to the mean control area deposition rate and expressing the ratio as a percent. The relative excess deposition rate of sulfate increased rapidly from $\sim 15\%$ at Site 18 to $\sim 25\%$ at Sites 19 and 25A then more slowly to $\sim 30-35\%$ at Sites 26 and 32, and to $\sim 45\%$ at Site 33. Although Site 18 was not directly influenced by plume presence during the storm, the plume remained over this site for 0.17 hr just prior to the event such that plume-derived material could have been present during rain initiation. It appears that there was a relationship between plume residence time and excess deposition rate such that excess sulfate deposition reached a "steady-state" value for plume residence times of ~ 1 to 2 hr. Although this relationship suggests a plume washout effect on sulfate deposition in the target area, analysis of this trend was complicated by variations in rainfall rate, which also increased as plume residence

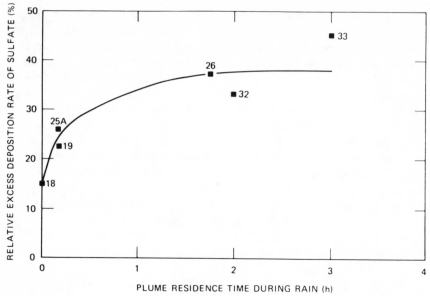

Figure 4. Characteristics of plume washout of sulfate as a function of the plume residence time during rain for the storm of December 13, 1979. Relative excess plume washout is calculated as the ratio between the excess deposition rate in the target area (target area deposition rate at each site minus the average control area deposition rate) and the mean deposition rate in the control area expressed as a percent.

time increased. The relative importance of each factor will be difficult to determine without further field studies in which a sufficient number of storms must be sampled to increase the chances of sampling several events of uniform rainfall rate over the network.

ACKNOWLEDGMENTS

This work was performed as part of the program on the Meteorological Effects of Thermal Energy Releases, sponsored by the Division of Advanced Nuclear Systems and Projects and the Office of Health and Environmental Research, U.S. Department of Energy under contract W-7405-eng-26 with the Union Carbide Corporation–Nuclear Division. By acceptance of this chapter, the publisher or recipient acknowledges the U.S. Government's right to retain a nonexclusive, royalty-free license in and to any copyright covering the chapter. The authors are indebted to A. A. N. Patrinos, L. Jung and R. L. Miller of Oak Ridge National Laboratory for their help in sampling. The advice of R. de Pena (Pennsylvania State University) and R. Turner

(Environmental Sciences Division, Oak Ridge National Laboratory) on sampling strategy and data analysis is gratefully acknowledged. Publication No. 1592, Environmental Sciences Division, Oak Ridge National Laboratory.

REFERENCES

1. Patrinos, A. A. N., and H. W. Hoffman. "Meteorological Effects of Thermal Energy Releases (METER) Program Annual Progress Report October 1978 to September 1979," ORNL/TM-7286, Oak Ridge National Laboratory, Oak Ridge, TN (1978).
2. Chen, N. C. J., and A. A. N. Patrinos. "Feasibility of Conducting Wetfall Chemistry Investigations Around the Bowen Power Plant," ORNL/TM-6930, Oak Ridge National Laboratory, Oak Ridge, TN (1979).
3. Högström, U. *Atmos. Environ.* 8:1291-1303 (1974).
4. Hutcheson, M. R., and F. P. Hall. *Atmos. Environ.* 8:23-28 (1974).
5. Dana, M. T., J. M. Hales and M. A. Wolf. *J. Geophy. Res.* 80:4119-4129 (1975).
6. Li, T. Y., and H. E. Landsberg. *Atmos. Environ.* 9:81-88 (1975).
7. Wiebe, H. A., and D. M. Whelpdale. "Precipitation Scavenging from a Tall-Stack Plume," in *Precipitation Scavenging* (1974), CONF-741003, National Technical Information Service, Springfield, VA (1977).
8. Marsh, A. R. W. *Atmos. Environ.* 12:401-406 (1978).
9. Lindberg, S. E., et al. "Mechanisms and Rates of Atmospheric Deposition of Selected Trace Elements and Sulfate to a Deciduous Forest Watershed," ORNL/TM-6674, Oak Ridge National Laboratory, Oak Ridge, TN (1979), 514 pp.
10. Miller, R. L., R. E. Saylor and A. A. N. Patrinos. "The METER-ORNL Precipitation Network from Design to Data Analysis," ORNL/TM-6523, Oak Ridge National Laboratory, Oak Ridge, TN (1978).
11. Volchok, H., L. Toonkel and M. Schonberg. "Trace Metals: Fallout in New York City," HASL-281, Health and Safety Laboratory, NY (1974).
12. Lindberg, S. E., et al. "Walker Branch Watershed Element Cycling Studies: Collection and Analysis of Wetfall for Trace Elements and Sulfate," in *Watershed Research in Eastern North America*, D. L. Correll (ed.) (Edgewater, MD: Smithsonian Institute Press, 1977), pp. 125-150.
13. McSwain, M. R., R. J. Watrous and J. E. Douglass. *Anal. Chem.* 46:1329-1331 (1974).
14. Gatz, D. and A. N. Dingle. *Tellus.* 23:14-27 (1971).
15. Raynor, G., and J. Haynes. "Experimental Data from Analysis of Sequential Precipitation Samples at Brookhaven National Laboratory," BNL-50826, Brookhaven National Laboratory, NY (1977).
16. Panofsky, H. A., and G. W. Brier. *Some Applications of Statistics to Meteorology*, The Pennsylvania State University, University Park, PA (1968), pp. 147-153.
17. Patrinos, A. A. N., N. C. J. Chen and R. L. Miller. *J. Appl. Meteor.* 18:719-732 (1979).

18. Schlesinger, W. H., W. Reiners and D. Knopman. *Environ. Pollut.* 6: 39–47 (1974).
19. Struempler, A. W., *Atmos. Environ.* 10:33–37 (1976).
20. Pack, D. H. *Science,* 208:1143–45 (1980).
21. Stensland, G. J. "Numerical Simulation of the Washout of Hygroscopic Particles in the Atmosphere," PhD thesis, The Pennsylvania State University, University Park, PA (1973).

CHAPTER 3

SOME EVIDENCE FOR ACID RAIN RESULTING FROM HCl FORMATION DURING AN EXPENDABLE VEHICLE LAUNCH FROM KENNEDY SPACE CENTER

B. C. Madsen

Department of Chemistry
University of Central Florida
Orlando, Florida 32816

INTRODUCTION

During the past 20 years Kennedy Space Center (KSC), Florida, has been the site of both manned and unmanned expendable vehicle launches (EVL). A solid propellant booster system was used during the launching of Delta and Titan III rockets. The space shuttle has an even larger solid propellant booster system than either the Deltas or Titan IIIs.

A significant portion of solid rocket motor (SRM) exhaust generated during initial ignition and liftoff of a launch vehicle remains in the vicinity of the launch pad and forms a ground cloud after mixing with air. The ground cloud cools by adiabatic expansion and further mixing with air as it rises. Ground cloud stabilization typically occurs below 5-km altitude depending on atmospheric conditions. Diffusion and downwind drift of the ground cloud ultimately dissipate the high concentrations of exhaust components.

Amounts of propellant utilized and some ground cloud characteristics for various solid rocket motors are summarized in Table 1. Burning of the solid propellant by various launch vehicles generates $\sim 21\%$ by mass of HCl as exhaust product [2]. Should an overriding rainshower or possible exhaust-cloud-induced rain occur before dissipation of the ground cloud, then acid

Table 1. Propellant and Exhaust Characteristics for SRM Launches

	Shuttle	Titan III	Delta (Caster IV)
SRM Propellant (kg)	1,080,000	430,900	81,600
Propellant consumed below 5 km (kg)	317,500	142,000	26,000
HCl formed below 5 km (kg)	63,500	29,900	5,400
Ground cloud stabilization altitude (km)	0.8-5[a]	1-2	
HCl in stabilized ground cloud (nominal max ppm)	5-40[a]	5-40	
HCl in stabilized ground cloud (kg)	32,200 [a]	15,000	

[a]Predicted (Ref. 1).

rain is likely. The occurrence of rain during or immediately after launch may lead to scavenging of HCl and result in acid rain. Predictions [2] that are based on a multilayer diffusion model which was summarized by Susko [3] indicate that acid rain could occur following launch of the shuttle if extreme conditions in launch cloud dynamics and local meteorology exist. Worst-case isolated rain showers with pH of <1.0 could occur within 25 km of the launch site and less acidic showers could occur at extended distances following Titan III or shuttle launches. Delta SRMs are considerably smaller than either Titan III or shuttle SRMs. Delta SRM exhaust cloud characteristics have not been evaluated. Total HCl formation below 5 km from Delta launches is considerably less than that from other SRM launches. A less severe acidic rain potential should exist based on comparison data which are available as summarized in Table 1.

A program designed to characterize the composition of rain in the vicinity of KSC during the period immediately prior to launch of the shuttle is in progress [4]. Annual weighted mean rainfall pH of ~ 4.6 has been measured during the past three years. The presence of sulfuric acid and nitric acid appears to be exclusively responsible for the observed acidity, and the presence of chloride in precipitation is attributed to the marine environment. The elemental Cl/Na ratio is slightly below that present in seawater. The presence of acidity due to HCl in rain associated with launches that utilize SRMs should be characterized by chloride concentrations in excess of those contributed by sea salt. Any excess can be calculated as follows:

$$\text{Excess } Cl^- = [\text{Total } Cl^-] - 1.16\,[Na^+]$$

where [Total Cl⁻] and [Na⁺] are experimentally measured concentrations and 1.16 is the theoretical Cl/Na (eq/eq) ratio for seawater.

During three Titan III EVL from KSC in 1975, 1977 and 1978, a special NASA task force collected droplets of rainwater [5], which, on chemical analysis, showed chloride concentrations in excess of those expected to be of marine origin. Acidity of these droplets was not measured; however, at some locations pH paper was spotted with very acidic droplets.

On February 14, 1980, a Delta EVL at launch complex 17 carried a Solar Maximum Mission Observatory into earth orbit. The caster IV SRM configuration generates about 5400 kg of HCl during ignition, liftoff and the initial 5-km climb. Shortly after this launch, areas adjacent to the launch pad were covered by a rain shower which deposited ~ 0.15 cm of precipitation. Evidence is presented that shows that the unusual acidity of rain collected at one site was due to the presence of hydrochloric acid in the samples collected.

EXPERIMENTAL

Permanent (P) and temporary (T) sites identified in Figure 1 were used to collect samples. Permanent rainfall collection sites were equipped with Aerochem Metrics Model 201 (HASL-type) collectors. Rainfall collections including those during the launch-associated rain were made in buckets with a 640-cm² collection surface. Clean buckets were deployed at ground level at temporary sites T1-T5 one hour prior to launch. Two HASL-type collectors (T5a and T5b) were also located at site T5. These collectors had been deployed with collection buckets in place 72 hours prior to launch and were activated 24 hours before launch. These two collectors received a small amount of rain prior to 9:00 a.m. on February 14, 1980, in addition to the launch-associated rain. All launch-associated samples were collected $\sim$2-hr post launch after the rain had ceased. Samples were analyzed using methods described previously [4].

RESULTS AND DISCUSSION

Results presented in Table 2 summarize the composition of rain that occurred most recently before and after the launch-associated rain, and detail the composition of the launch-associated rain. Rains that occurred on February 10, 1980, and late on February 14, 1980, were characterized by moderate acidity. The concentrations of nitrate and excess sulfate (sulfate in excess of that of marine origin) were great enough to account for the

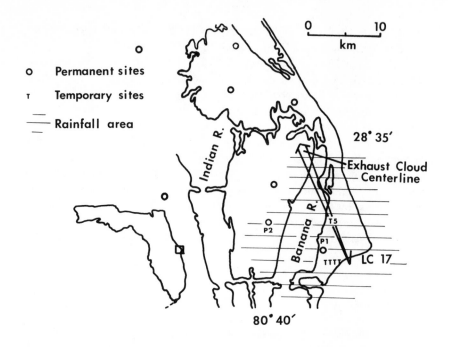

Figure 1. Rain coverage and sampling locations.

measured acidity. In addition, the Cl/Na ratio for samples collected during these two rains was essentially that which is to be expected if all chloride and sodium are of marine origin (1.16 eq/eq). Concentrations of nitrate and excess sulfate in launch-associated rain samples were great enough to account for determined acidity for the sites identified as P1, P2 and T1-T4. In addition the Cl/Na ratio in samples from these sites was somewhat less than the expected ratio for marine-derived components except at site P2. The sample collected at site P2 had a small excess Cl⁻ concentration. It can be postulated that all chloride in samples which contained no excess chloride was of marine origin. The three launch-associated samples collected at site T5 yielded higher acidities which could not be accounted for by nitrate and excess sulfate. They also exhibited elevated Cl/Na ratios, which indicate the presence of excess chloride. Anion/cation (eq/eq) ratios determined from measured concentrations of all major anions and cations [4] present in the samples ranged from 0.94 to 1.06. The Mg/Na (eq/eq) ratio for individual samples ranged from 0.20 to 0.24. The Mg/Na (eq/eq) ratio is expected to be 0.19 if the presence of magnesium and sodium is exclusively of marine origin.

Table 2. Rainfall Composition Before, Associated With, and After an Expendable Vehicle Launch[a]

Event		cm	pH	H	Na	Cl	Excess Cl	NO$_3$	Excess SO$_4$	Cl/Na
						(microequivalents/liter)				
2/10/80		0.48	4.87	13.6	25.4	28.2	0.0	8.7	11.2	1.11
		(0.26)		(2.9)	(4.7)	(5.3)		(4.3)	(3.3)	(0.09)
Launch-associated rain	P1	0.26	4.25	56.2	36.9	36.0	0.0	28.0	39.2	0.97
	P2	0.22	4.18	66.0	38.2	49.3	5.0	34.3	38.5	1.28
	T1	0.13	4.42	38.0	35.2	33.5	0.0	24.5	26.7	0.95
	T2	0.13	4.40	39.8	35.2	32.1	0.0	23.7	27.1	0.92
	T3	0.14	4.38	41.6	34.7	32.7	0.0	30.3	29.3	0.94
	T4	0.14	4.45	35.4	32.6	34.9	0.0	30.8	26.8	1.07
	T5	0.17	3.87	135.0	42.1	116.7	67.6	30.3	48.4	2.76
	T5a	0.14	3.98	105.0	50.4	80.6	21.9	36.4	59.1	1.59
	T5b	0.15	3.91	123.0	94.7	151.9	41.6	57.2	58.0	1.60
2/14/80		1.34	4.45	35.4	13.1	15.2	0.0	11.9	27.3	1.16
		(0.26)		(4.6)	(4.0)	(4.3)		(2.4)	(3.7)	(0.16)

[a]Values in parentheses are standard deviations for cm and ratios, and precipitation weighted standard deviation for the ionic concentrations (see Ref. 6). Data were obtained from eight permanent sites.

The composition of rain samples collected at site T5 is not typical of composition of rain at KSC. The mean composition of rain that occurred at KSC during February for the past three years and at site P1 during February 1980 is summarized in Table 3. These results—which are consistent with those previously reported [4]—allow comparisons and conclusions to be made about measured acidity, marine-derived components, nitrate and excess sulfate. Specifically, nitrate plus excess sulfate concentrations are great enough to more than account for measured acidity, the Cl/Na ratios do not indicate presence of excess chloride, and the Mg/Na ratios reinforce the assumption that sodium concentrations are indeed due to the marine influence.

Although the presence of hydrochloric acid is apparent, the inconsistency in results for the three samples collected at site T5 is not easily understood. A probable explanation is contamination of samples T5a and T5b with sea salt mist or aerosol. Sample T5 was collected in an open bucket that was deployed ~1 hr prior to launch and was picked up ~2 hr after launch. Samples T5a and T5b included the launch-associated rain plus a small amount of rain from an early morning shower. Proximity of the site to the ocean (4 km) makes sea salt aerosol contamination a distinct possibility.

Evidence presented supports the conclusion that increased acidity in samples collected at site T5 is due to the presence of hydrochloric acid. Chloride concentrations in excess of those predicted from measured sodium concentrations which are assumed to be exclusively of marine origin are high

Table 3. Mean Rainfall Composition at KSC During February 1978-1980 and at Site P1 During February 1980

	1978	1979	1980	1980
Number of Sites	8	14	8	Site P1
Total Samples	48	47	43	7
cm	11.48	3.84	9.39	12.67
pH	4.41	4.46	4.59	4.69
H (μeq/L)	39.7	27.5	25.7	20.4
Na (μeq/L)	40.1	62.6	46.9	42.6
Cl (μeq/L)	44.5	70.6	52.2	47.2
NO_3 (μeq/L)	9.0	13.0	8.0	5.6
Excess SO_4 (μeq/L)	32.0	30.7	20.8	17.0
Cl/Na (eq/eq)	1.11	1.13	1.11	1.11
Mg/Na (eq/eq)	0.30	0.24	0.20	.19
Anion/Cation (eq/eq)	0.88	0.94	0.92	0.94

enough to support this conclusion. The general quality associated with sample analysis results is high, as shown by acceptable anion/cation ratios.

Evaluation of the composition of rain which occurred due to meteorological conditions that were present shortly after a SRM launch has shown that scavenging of HCl from the SRM exhaust does occur and leads to increased acidity. This acidic rain was apparently spatially isolated and less acidic than several previously studied rains [4] which occurred in the absence of SRM exhaust.

ACKNOWLEDGMENTS

The results presented here represent a portion of the work supported by NASA contracts NAS10-8986 and NAS10-9841. The participation of John Hogsett, who collected and analyzed the samples, is appreciated.

REFERENCES

1. "Environmental Impact Statement, Space Shuttle Program—Final Statement," NASA Headquarters, Washington, DC (April 1978).
2. Pellett, G. L. *ERDA Symposium Series,* Precipitation Scavenging Conference, 41:437–65 (1977).
3. Susko, M. *J. Appl. Meteorol.* 18:48 (1979).
4. Madsen, B. C. *Atmos. Environ.* 15:853 (1981).
5. Pellett, G. L., et al. 73rd Annual Air Pollution Control Association Meeting. Paper No. 80–49.6 (1980).
6. Liljestrand, H. M., and J. J. Morgan. *Tellus* 31:421 (1979).

PART 2

REGIONAL EFFECTS

CHAPTER 4

DEPOSITION AND TRANSPORT OF HEAVY METALS IN THREE LAKE BASINS AFFECTED BY ACID PRECIPITATION IN THE ADIRONDACK MOUNTAINS, NEW YORK

David E. Troutman

U.S. Geological Survey
Water Resources Division
Tallahassee, Florida 32303

Norman E. Peters

U.S. Geological Survey
Water Resources Division
Albany, New York 12201

INTRODUCTION

Elevated concentrations of heavy metals in lake waters and sediments in regions of North America and Scandinavia receiving acid atmospheric precipitation have been reported in recent years [1-3]. Increases over the last 30 years in the concentration of heavy metals in dated cores of lake sediments from New England coincide with reported increases in atmospheric deposition of sulfates and nitrogen oxides, the strong-acid components of atmospheric precipitation [3, 4]. Continued increases in the bioaccumulation of some heavy metals could lead to health hazards as well as sustained environmental damage.

These elevated concentrations of heavy metals in remote lakes have been attributed to their increased concentration in atmospheric deposition and their increased mobilization from the basins or lake sediments. Increased

33

mobilization is hypothesized to result from increased solubility of these metals in waters of low pH [5, 6].

This report examines the hypothesis that the release of iron, lead, manganese and zinc within lake basins receiving acid atmospheric precipitation is a function of the extent of acid neutralization. To test this hypothesis, three lake basins were chosen in the Adirondack Mountain region of New York State (Figure 1), an area sensitive to acid precipitation [7].

The lake watersheds were selected such that one (Woods Lake) would be acid (pH $\cong$ 4-5), a second (Sagamore Lake) would be intermediate in acidity (pH $\cong$ 5-6) and a third (Panther Lake) would be neutral (pH $\cong$ 6-7) [8]. Additional criteria in lake selection were that all basins be close to each other, geologically similar, relatively remote and undisturbed, accessible year round, of similar size, relief and elevation, and have gaugeable outlet streams and minimal seepage loss. A watershed of intermediate pH conforming to all criteria could not be located; the one selected (Sagamore Lake) is significantly larger than the others but suitable in all other aspects. The acid and neutral basins conform to all criteria. The study was conducted for one year from June 1, 1978 to May 31, 1979.

BASIN CHARACTERISTICS

Physiography and Geology

The three lake basins studied are within 25 km of one another in the southwestern region of the Adirondack Mountains. Runoff from each basin discharges into a lake immediately above the basin outlet. The physical and chemical characteristics of the lakes and basins are summarized in Table 1.

All three basins are forested except in areas of exposed bedrock. Bogs and wetlands occupy a significant part of the upland areas of the larger (intermediate pH) basin. Deciduous forest cover predominates in all three basins and consists of second-growth hardwoods—*Fagus grandifolia* (American beech), *Betula alleghaniensis* (yellow birch), *Acer saccharum* (sugar maple) and *Acer rubrum* (red maple). Spruce, fir and herbaceous vegetation occupy the lakeshores and bogs [9].

All three basins are of similar crystalline bedrock type that is resistant to weathering. Bedrock in the acid basin consists of a hornblende granitic (charnokitic) gneiss with minor outcrops of pink granitic and syenitic gneiss. Bedrock in the neutral basin has consistent outcropping of biotite granitic gneiss, and bedrock in the intermediate basin is comparable to that of the acid basin except that it contains considerably more amphibolitic layering.

Mineralogic and chemical analyses of bedrock from the three basins are

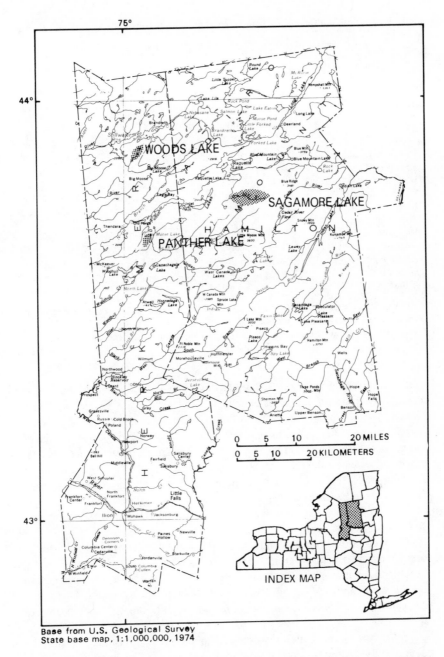

Base from U.S. Geological Survey
State base map, 1:1,000,000, 1974

Figure 1. Location of Woods (acid), Sagamore (intermediate) and Panther (neutral) Lakes, New York.

Table 1. Physical and Chemical Characteristics of Acid, Intermediate and Neutral Basins

Characteristic	Basin		
	Acid	Intermediate	Neutral
Watershed area (km^2)	2.07	49.7	1.24
Relief (m)	122	561	174
Forest Cover (%)	100	97	100
Lake Data:			
Elevation (m above sea level)	606	580	557
Area (km^2)	0.23	0.72	0.18
Watershed ratio	0.11	0.015	0.15
Water volume (m^3)	8.13×10^5	75.4×10^5	7.09×10^5
Residence time (days)	221	81.4	264
Mean depth (m)	3.5	10.5	3.9
Maximum depth (m)	12.0	23.0	7.0
Mean alkalinity[a] (μeq/L)	–30	10	100
Alkalinity range (μeq/L)	66 to –44	103 to –2	220 to –38
Mean pH[b]	4.6	5.4	5.7
pH range	5.0 to 4.4	6.8 to 4.9	7.5 to 4.8

[a]Negative alkalinities (strong acidities) calculated from total hydrogen-ion concentration as determined by base titration to pH 4.5.
[b]Mean pH calculated based on flow-weighted hydrogen-ion concentration.

presented in Tables 2 and 3, respectively. The predominant minerals that weather by hydrogen-ion transfer with mineral cations are plagioclase, hornblende, pyroxenes and biotite. Volume percentages determined by petrographic point counts of thin sections indicate that, although bedrock in all basins is relatively unreactive, bedrock in the neutral basin is the least reactive, followed by the intermediate, then acid basins.

Comparison of mineral and chemical composition of bedrock samples from all three basins (Tables 2 and 3) suggests that higher weight percentages of iron, titanium, magnesium and calcium are associated with higher percentages of hornblende.

The amphibolitic (hornblende-rich) layers of bedrock in the intermediate basin range in thickness from a few mm to about 1 m and contain mostly hornblende. Although these layers seem to make up only a small percentage of the total bedrock, in all observations they had a lower relief than the adjacent leucocratic layers in outcrops, which indicates a higher degree of weathering.

All three basins contain varying amounts of unconsolidated surficial deposits; these deposits consist primarily of sandy till. Thickness of till ranges from <3 m at higher elevations and on steeper slopes to >10 m in some

Table 2. Mineral Composition of Bedrock, Surficial Material and Lake Sediments of Acid, Intermediate and Neutral Basins[a]

Bedrock — Volume percent[b]

Mineral	Acid	Intermediate	Neutral
Quartz	26	29	37
Plagioclase	14	12	11
K-feldspar	46	43	51
Hornblende	8	11	ND
Orthopyroxene	3	ND	ND
Clinopyroxene	1	1	ND
Biotite	1	2	2
Apatite	tr	tr	tr
Opaques	tr	1	tr
Number of Samples	7	13	5

Surficial Material (Whole rock[c] and <2 micrometers) and **Sediments** (Whole rock[c], d and <2 micrometers)

Mineral	Surficial Acid	Surficial Intermediate	Surficial Neutral	Sediments Acid	Sediments Intermediate
Whole rock					
Quartz	a	a	a	a	a
Plagioclase	a	a	a	a	a
K-feldspar	m	m	a	m	m
Amphibole	m	m	m	tr	a
<2 micrometers					
Illite	x	x	x	x	x
Chlorite	x	x	x	x	x
Mixed layer	x	x	x	x	x
Amphibole	x	x	x	ND	ND
Talc	x	x	x	ND	ND
Hydrobiottite	ND	ND	x	ND	ND
Vermiculite	x	x	x	x	x
Number of Samples	4	5	3	3	3

[a] a = abundant; m = minor; tr = trace; x = present; ND = not detectable.
[b] Volume percentages determined through thin-section point counting.
[c] Quantitative mineral content determined through X-ray diffraction techniques of both whole rock and the <2-μm mineral fractions.
[d] No samples analyzed from neutral lake.

Table 3. Mean Chemical Composition and Standard Deviation of Bedrock, Surficial Material and Lake Sediments of Acid, Intermediate and Neutral Basins, in Oxide Weight Percent[a]

Component	Bedrock			Surficial Materials			Sediments		
	Acid	Intermediate	Neutral	Acid	Intermediate	Neutral[b]	Acid	Intermediate	Neutral
SiO_2	64.86±6.30	66.20±12.30	75.51±1.71	70.41±6.10	71.01±1.66	73.14±1.54	50.80±6.21	48.53±1.09	37.38±1.61
TiO_2	0.53±0.09	0.98±1.17	0.19±0.10	0.80±0.35	0.91±0.03	0.83±0.11	0.06±0.00	0.52±0.03	0.14±0.03
Al_2O_3	15.67±3.18	13.21±1.76	13.01±1.44	11.80±1.56	12.58±0.76	10.53±1.01	5.87±1.30	9.68±0.63	5.01±0.68
FeO	6.78±1.75	7.37±6.79	1.51±0.64	4.13±1.46	4.64±0.58	4.72±0.78	1.95±0.46	3.97±0.76	3.00±0.55
MnO	0±0	0.02±0.03	0±0	0.03±0.07	0±0	0±0	0±0	0.03±0.05	0.21±0.02
MgO	0.49±0.26	1.50±2.29	0.06±0.08	0.47±0.12	0.90±0.19	1.19±0.23	0.10±0.02	0.64±0.05	0.15±0.06
CaO	2.74±1.14	3.12±2.96	0.86±0.25	1.39±0.34	2.10±0.31	1.99±0.40	0.37±0.10	1.74±0.11	0.67±0.06
Na_2O	4.12±1.42	3.36±0.75	3.54±0.35	2.57±0.04	2.98±0.35	2.36±0.33	0.16±0.08	1.64±0.13	0.06±0.06
K_2O	4.35±1.39	3.82±1.72	4.72±0.64	3.74±0.40	3.50±0.32	2.97±0.17	0.14±0.02	1.55±0.10	0.36±0.07
LOI	0.44±0.18	0.40±0.24	0.59±0.03	4.64±3.42	1.37±0.38	2.26±1.13	40.54±8.10	31.69±0.50	53.00±2.13
Number of Samples	4	7	3	4	4	3	3	3	10

[a]Oxide percentages were determined on fused samples using lithium tetraborate flux and analyzed by electron microprobe. All analyses have been normalized.

[b]Determined by X-ray fluorescence.

lowlands. Depths of unconsolidated material, as inferred from seismic-refraction transects, are > 10 m in 5-10% of the acid basin; undetermined in most of the intermediate basin (because of its size and inaccessibility), and > 12 m in more than 50% of the neutral basin. Topography of the neutral basin suggests that thick glacial deposits (thicker than 10 m) occupy a percentage of basin area intermediate between those of the acid and neutral lakes.

In summary, the major geologic differences among the three basins is the percentage of basin area occupied by thick till. The neutral basin contains the greatest percentage, the acid basin the least.

X-Ray diffraction of the unconsolidated material and lake sediments indicates no significant mineralogic differences among the three basins (Table 2). Quantitative differences in the mineral composition of surficial material and lake sediments may be present, but the information is inconclusive. Heavy-mineral analysis of soils and surficial material from the basins indicates a greater abundance of hornblende in the intermediate basin (6.2 wt %) than in the acid basin (2.6 wt %) or neutral basin (2.7 wt %) [10].

Hydrology

During June 1, 1978, to May 31, 1979, the basin of intermediate pH received the smallest annual quantity of precipitation, 97.3 cm of water; the neutral basin received the most, 122 cm; and the acid basin received 118 cm. Precipitation had a weighted mean annual pH of 4.10 and was evenly distributed throughout the year. Approximately 40% of the total precipitation in each basin fell as snow, with accumulations slightly higher in the acid basin. Peak water equivalents of snowpack in each basin occurred during mid-February 1979; the values were: acid basin, 27 cm of water; intermediate basin, 16 cm; and neutral basin, 21 cm [10].

The annual water balance of each basin is summarized in Table 4. These values are similar to the 20-year averages for the Hubbard Brook Experi-

Table 4. Annual Water Balance of Acid, Intermediate and Neutral Basins—
June 1, 1978 to May 31, 1979[a]

	Acid	Intermediate	Neutral	HBEF[b]
Precipitation	118.0 (100)	97.3 (100)	122.0 (100)	130.0 (100)
Streamflow	65.6 (56)	68.9 (65)	79.5 (65)	80.1 (62)
Evapotranspiration (residual)	51.9 (44)	28.4 (29)	42.5 (35)	49.4 (38)

[a]First value is centimeters of water; number in parentheses is percentage of total.
[b]Hubbard Brook Experimental Forest, New Hampshire; 20-yr average from 1965 to 1974 (from Likens,et al. [14]).

mental Forest in the White Mountains of New Hampshire [7], which suggests that precipitation conditions were average during the collection period in this study area.

From hydrogeologic information provided by seismic profiles at the basin outlets, water loss from the basins by underflow at the outlets was determined to be insignificant [11]. Evapotranspiration from each basin was therefore calculated as the residual in the water balance.

The comparatively low annual quantity of precipitation measured in the intermediate basin may be a result of insufficient areal rainfall data from that basin. (The large area and inaccessibility of the upper part deterred collection of data at additional sites.) If a rainfall of 120 cm/yr, similar to that of the two other basins, is assumed, the calculated percentage of evapotranspiration in the water balance of the intermediate basin would be 43%, which approximates evapotranspiration values calculated for the other two basins.

Input loads reported herein for the intermediate basin are based on the actual measured quantity of precipitation (97.3 cm) in that basin. If the precipitation value is, in fact, underestimated, input loads calculated for that basin may be low by as much as 20%. However, even a 20% underestimation of input loads would not alter the conclusions and hypotheses proposed from this study.

Annual hydrographs and temporal trends in pH at the three basin outlets are illustrated in Figure 2. From data on the surficial geology of the basins and a comparison of the three annual hydrographs, significant differences in the hydrologic response and physical characteristics of the basins become apparent. The acid basin is a surface-water-dominated system because it has a thin cover of till that provides little storage capacity and thus produces rapid "flashy" runoff, characterized by a spiked hydrograph (Figure 2A).

In contrast, the neutral basin is a groundwater-dominated system in which flow to the lake is at a relatively uniform rate except during high runoff or melt periods. The extensive deep deposits of till at the lower elevations provide significant groundwater storage and seepage to the lake. Streams flowing into the neutral lake are intermittent and discharge only during periods of high rainfall and snowmelt. Comparison of the attenuated peaks in the neutral-lake hydrograph (Figure 2C) with the "flashy" acid-lake hydrograph (Figure 2A) illustrates this difference.

The intermediate basin is significantly larger than the other two and is characterized by extensive bogs in the upper elevations. The intermediate basin, much like the acid basin, exhibits a "flashy" response during runoff, but also has a sustained groundwater component that is derived from deep till and alluvial deposits in the lower basin (Figure 2B). Lost Brook, the inlet stream to the intermediate lake, drains 90% of the total drainage area but accounts for 97% of the total annual inflow of the lake.

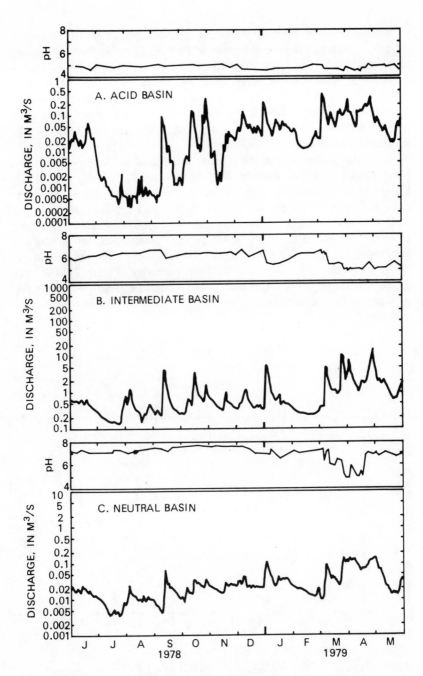

Figure 2. Annual hydrographs and variations in pH of lake outflows.

A comparison of the basins indicates that streamflows adjusted for areal differences among basins as illustrated in Figure 3 and pH differences among basins (Figure 2) are greatest during the periods of lowest flow (June–November). Adjusted low flows from the neutral lake were significantly greater than those from the acid lake and somewhat greater than those from the intermediate lake. The lack of inflow streams into the neutral lake implies that the difference in outflow from the basins results from differences in the respective groundwater regimes.

Biological productivity in the lakes of the neutral and intermediate basins is sufficient to cause increases in pH of these waters during the summer [12]. However, the comparably high pH of the lake outflows during February, when the lakes are biologically least active, suggests another control on pH.

Limited data collected from wells in the basins indicate that pH and alkalinity of groundwater in the till deposits of all basins are similar and groundwater is generally neutral (pH $\cong$ 7). In addition, the pH decrease at the lake outlets is accompanied by high discharges (Figure 2). The large differences in pH among the lake waters, therefore, seems to result predominantly from differences in the quantity of neutral groundwater flowing into the lakes.

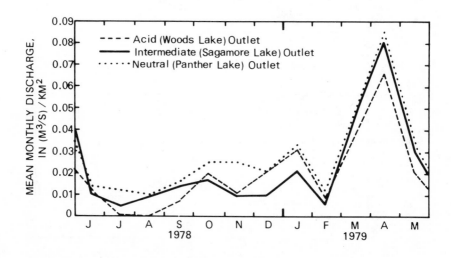

Figure 3. Mean monthly area-adjusted discharge at lake outflows from acid, intermediate and neutral basins.

Limnology

The neutral and acid lake basins have comparable physical and limnological characteristics, yet differ markedly in the extent to which acid precipitation is neutralized, as indicated in Table 1. All three lakes reach the lower pH limit almost simultaneously during the initial spring melt, when a large and continuous release of acid melt water occurs (Figure 3). Owing to lake stratification under the ice cover, only the top meter or so of the lake epilimnion, including parts of the epilimnion in contact with the littoral zone, is exposed to the highly acid melt waters so that the more alkaline water of the hypolimnion remains relatively unaffected. As the ice cover melts, however, the lakes quickly destratify, and the acid meltwater at the surface becomes mixed and diluted with the more neutral, larger volume of water of the hypolimnion [13]. The hypolimnion of all three lakes remains oxygenated throughout the year.

METHODS

Field Methods

Bulk precipitation (wetfall plus dryfall) samples for heavy-metals analysis were collected at approximately 6-week intervals from clearings within or adjacent to each basin. Because a mass-balance approach was taken, no attempt was made to differentiate between the wet and dry portion entering the basins. The collector was similar to that used in the Hubbard Brook watershed [14] and consisted of a 27-cm polyethylene funnel connected to a 4-L polyethylene receiving bottle. A vapor lock was placed in the silastic connecting tube to retard evaporation and prevent enrichment of chemical constituents in the sample. Precipitation quantity was determined from a network of seven weighing-bucket rain gauges. Two sites were established within or adjacent to each lake basin.

"Grab samples" of streamflow were collected for a period of one year (June 1978 to May 1979). Samples were collected biweekly during winter and summer baseflows and more frequently during periods of snowmelt and storm runoff. A stage-activated automatic sampler was used to collect samples during several storms during ice-free periods. All stream samples were taken from the centroid of flow directly above gauged sites at each lake outflow.

Streamflow was recorded continuously at each of the three lake outflows. The large drainage area of the intermediate lake basin (49.7 km^2), as compared with the acid basin (2.07 km^2) and neutral basin (1.24 km^2), necessitated installation of a stream gauge on Lost Brook, the inflow to the intermediate lake. Records for the Lost Brook gauge during January–March, when backwater resulted from ice, were estimated from lake-outlet records. Readings of lake elevation of all three lakes were collected weekly.

Analytical Methods

All precipitation and streamflow samples for heavy-metals analysis were collected in polyethylene containers that had been washed with concentrated nitric acid (HNO_3) to remove impurities and were rinsed three times with deionized water. Sample bottles were flushed with stream water before collection. All samples for iron, lead, zinc and manganese were acidified to pH 1.0 to 2.0 with ultrapure HNO_3 soon after collection.

Specific conductance and pH of all streamflow samples were determined in the field. Data on pH and specific conductance of precipitation were obtained from event samples [11]. (An event sample is defined as the amount of precipitation that has a water content in excess of 0.25 cm and that has fallen within a 24-hr period beginning at 6:00 a.m.) Analysis for heavy metals was made by the U.S. Geological Survey (USGS) Central Laboratory in Atlanta, Georgia, for total "recoverable" concentrations by atomic absorption spectrometry. Manganese, iron and zinc were aspirated directly into the air-acetylene flame; lead was aspirated following chelation with ammonium pyrrolidine dithiocarbamate (APDC) and extraction with methylisobutyl ketone (MIBK) [15]. Detection limits for iron, manganese and zinc were 10 μg/L and for lead was 1 μg/L. A program of quality assurance used by the USGS on all samples analyzed is described by Friedman et al. [16].

Data Analysis

Monthly input loads entering the basins were calculated as follows:

$$\text{Input} = \sum_{i=1}^{n} (Q_p C K_1)$$

where Input = monthly loads entering each basin from atmospheric deposition in (g/ha)/mo

Q_p = quantity of precipitation, in cm/mo, in each watershed for the part of each month in which the collector was exposed

C = concentration of total "recoverable" iron, lead, manganese or zinc in the bulk precipitation sample, in μg/L, as determined by laboratory analysis

K_1 = conversion factor

Monthly outflow loads transported from each basin were calculated as follows:

$$\text{Outflow} = \sum_{i=1}^{n} (\bar{Q}_S C t K_2)$$

where Outflow = monthly loads transported from each watershed, in (g/ha)/mo
$\bar{Q}_S$ = mean daily discharge, in m^3/sec, as recorded at the gauged lake outflows between sampling dates
C = concentration of total "recoverable" iron, lead, manganese or zinc, in $\mu g/L$, as determined by laboratory analysis of stream samples
t = number of days between successive samples
K_2 = conversion constant based on drainage area of the basin.

Monthly input and outflow loads were summed over the 12-month period of study to determine annual loads in (g/ha)/yr. Ratios of annual input to outflow (I/O) loads were calculated; ratios < 1 indicate a loss, or transport, of a constituent from the basin from weathering and ratios > 1 indicate a retention or accumulation of that constituent in the basin. Fluxes were calculated by determining the differences between outflow and input loads; a net transport of a constituent results if the difference is positive; a net accumulation results if the difference is negative.

Annual I/O loads in this study are estimated to be accurate within ± 20% of actual values. All USGS data collected from this study as well as information pertaining to dates and frequency of sample collection are published in USGS "Water Resources Data for New York, 1978 and 1979" [17].

RESULTS AND DISCUSSION

Iron

Monthly input and outflow loads of iron in the acid, intermediate and neutral basins during the period of study are depicted in Figure 4. From data collected over the 12 months of study, no significant relationship was discernable between neutralization of acid precipitation in the basins and iron release from the three watersheds. The ratios of I/O loads for the acid basin (2.2) and neutral basin (1.2) are > 1.0, which indicate a net accumulation (gain) of iron in both basins; the ratio for the intermediate basin (0.68) is < 1.0, which indicates a net transport (loss) of iron from that basin.

Atmospheric deposition of iron in all three basins had considerable monthly variability, as indicated by the means and ranges in concentration (Table 5 and Figure 4) and did not exhibit a consistent seasonal trend. Maximum

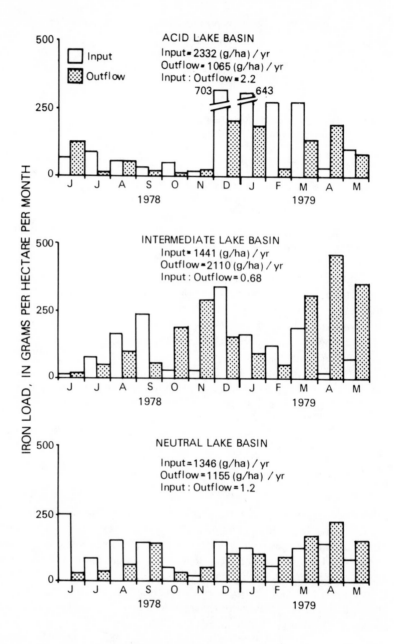

Figure 4. Monthly input and outflow loads of iron from acid, intermediate and neutral basins. Unshaded bars indicate input; shaded bars indicate outflow.

Table 5. Summary of Iron Data from Acid, Intermediate and Neutral Basins

	Streamflow				Precipitation		
	Acid	Intermediate	Lost Brook[a]	Neutral	Acid	Intermediate	Neutral
Mean concentration ± standard error (μg/L)	252 ± 57	299 ± 19	343 ± 69	286 ± 38	97.8 ± 45	121 ± 32	121 ± 17
Range in concentration (μg/L)	20–1500	40–750	110–2000	40–1200	10–440	10–300	10–340
Number of samples	42	42	29	52	9	9	8
Total annual flux (outflow minus inflow) (g/ha)	1065	2110	2240	1155	2332	1441	1346

[a]Lost Brook is inflow to neutral lake.

concentrations of 440 $\mu g/L$ in the acid basin and 300 $\mu g/L$ in the intermediate basin occurred in December 1978 and January 1979, but the maximum in the neutral basin (340 $\mu g/L$) occurred in June 1978. No relationship was found between monthly concentrations of iron in bulk precipitation samples and quantity of precipitation. The measured annual input loads of 2332 (g/ha)/yr in the acid basin, 1441 (g/ha)/yr in the intermediate basin, and 1346 (g/ha)/yr in the neutral basin in bulk collections (wet plus dry) are significantly higher than the 432 (g/ha)/yr input load calculated by Lazarus et al. [18] from 6 months of wetfall data collected in Albany, New York, 150-km southeast of the study basins. Pierson et al. [19] hypothesized that atmospheric concentrations of iron are derived primarily from the soil; therefore, the relatively high input loads measured in the three lake basins (relative to Lazarus' Albany data) suggest that a significant part of atmospheric deposition of iron probably occurs as dry deposition. If this is true, the differences in monthly deposition of iron among the three watersheds are probably a result of local influences (mining, construction, road dust, exudate from vegetation, etc.) rather than regional influences.

Iron concentrations in streamflow samples also showed considerable temporal variation and range (Table 5). Monthly outflow loads during summer and fall 1978, when snowpack storage and release is not a factor, correlate well with input loads (Figure 4). The spring melt (March–May) accounted for 38% of the annual outflow of iron from the acid basin, 46% from the neutral basin and 53% from the intermediate basin (Table 6).

Additional sources of iron transported from the basins are: (1) weathering of minerals containing iron, including chlorite, biotite and hornblende, by proton-transfer reactions; (2) reduction of iron in oxides and hydroxides and subsequent complexation with organic matter in reducing soils and bogs; (3) iron oxide and hydroxide coatings on suspended sediment and (4) plant exudate. Although hornblende is more prevalent in the weathering environment of the intermediate basin than the others and could theoretically provide more iron for weathering, the iron transported from the intermediate basin (2110 (g/ha)/yr) compared to the acid basin (1065 (g/ha)/yr) and neutral basin (1155 (g/ha)/yr) probably results more from the mobilization and complexation with organic matter in the bogs, which are intermittently flushed with water from snowmelt and storms.

Outflow loads of iron in the neutral basin showed less monthly variability than the other basins (Figure 4). The relatively small range in variability of monthly outflow loads of iron in that basin suggests a possible correlation of iron transport with basin hydrology; for example, the larger groundwater component in discharge from the neutral basin may cause iron transport from this basin to be more evenly distributed over a range of flows here than in other basins.

Table 6. Seasonal Percentage of Annual Iron (Fe), Lead (Pb), Manganese (Mn) and Zinc (Zn) Transport, and Discharge (Q) from Acid, Intermediate and Neutral Basins, 1978–1979

Season	Acid (%)					Intermediate (%)					Neutral (%)				
	Fe	Pb	Mn	Zn	Q	Fe	Pb	Mn	Zn	Q	Fe	Pb	Mn	Zn	Q
Summer 1977 (June–August)	18	24	3	3	5	7	24	4	3	9	10	23	7	7	10
Fall 1977 (September–November)	4	12	16	16	15	26	19	20	11	16	19	27	16	15	17
Winter 1977–1978 (December–February)	40	18	25	18	25	14	5	9	8	14	25	13	9	17	22
Spring 1978 (March–May)	38	47	56	62	54	53	51	67	78	60	46	37	68	61	51

Lead

Ratios of atmospheric inputs to outflows of lead for the acid, intermediate and neutral basins, as depicted in Figure 5, were 6.4, 3.4 and 3.2 respectively, which indicates a net accumulation in all three basins. In the acid basin, 183 (g/ha)/yr or 84% of the total input load, was retained; in the intermediate basin 116 (g/ha)/yr, or 70%, was retained; and in the neutral basin 115 (g/ha)/yr, or 69%, was retained. Nearly 150% or 68 (g/ha)/yr, more lead accumulated in the acid basin than in either of the other basins.

Input concentrations of lead in precipitation in all basins were highest in fall and summer and lowest during winter, which suggests: (1) a possible correlation with automobile use in the area, (2) more effective scavenging of lead by rain than by snow, or (3) a correlation with long-range transport resulting from air masses in the industrial Midwest [20]. A similar trend of higher summer concentrations was observed in seasonal deposition of lead at Hubbard Brook [21]. No significant correlation was found between precipitation quantity and lead concentrations in bulk samples. The range in concentration of bulk samples among basins was small (5-36 $\mu g/L$); the mean lead concentration was similar for all basins (Table 7).

In a similar mass-balance study of lead flux in a terrestial watershed at Hubbard Brook, Siccama and Smith [21] reported annual inputs of 317 (g/ha)/yr and accumulations of 305 (g/ha)/yr, whereas the mean annual input and accumulation of lead in the three Adirondack basins was 183 (g/ha)/yr and 138 (g/ha)/yr, respectively. The high accumulations in forested basins are believed to result from complexing of lead with organic matter and, in particular, sulfur and phosphate organic ligands or adsorption on iron and manganese oxides in the upper soil horizons [6, 22, 23]. Although the lead analyses were performed on the total sample in this study, it is probable that a significant part of this is in the dissolved form (<0.45 μm) correlative to observations at Hubbard Brook [21].

Concentrations of lead in streamflow samples were low compared to those in bulk precipitation samples and were often undetectable (<1.0 $\mu g/L$). The majority of lead transport from the three basins occurred during the melt period from March to May 1979, when 47, 51 and 37% of the total annual outflow load was transported from the acid, intermediate and neutral lake basins, respectively (Table 6). Monthly outflow loads of lead, like iron, showed less variability in the neutral basin than in the more acid basins (Figure 5).

Comparison of outflow minus input loads from the neutral and acid basins during summer and fall (June-November), when hydrologic regimes and groundwater flow of the two basins differed most, indicate that 44 g/ha accumulated in the neutral basin compared with 107 g/ha in the acid basin.

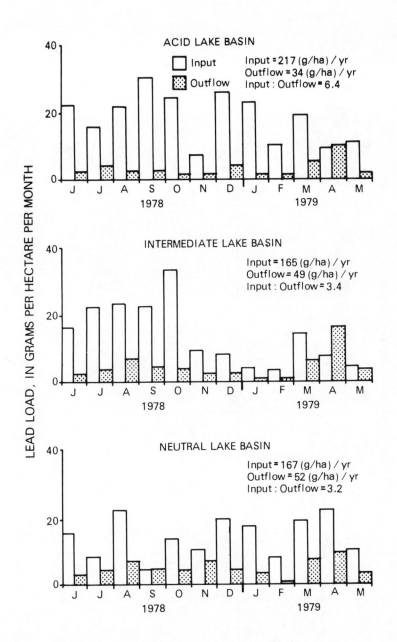

Figure 5. Monthly input and outflow loads of lead from acid, intermediate and neutral basins. Unshaded bars indicate input; shaded bars indicate outflow.

Table 7. Summary of Lead Data from Acid, Intermediate and Neutral Basins

	Streamflow				Precipitation		
	Acid	Intermediate	Lost Brook[a]	Neutral	Acid	Intermediate	Neutral
Mean concentration ± standard error (μg/L)	12 ± 2.7	6.2 ± 1.2	3.50 ± 0.8	12 ± 1.7	15 ± 2.6	18 ± 3.8	17 ± 8.7
Range in concentration (μg/L)	0–71	0–36	0–20	0–42	5–26	6–36	9–25
Number of samples	43	40	29	50	9	9	8
Total annual flux (outflow minus inflow) (g/ha)	34	49	25	52	217	165	167

[a]Lost Brook is inflow to neutral lake.

The difference in accumulated summer and fall loads between the two basins (63 g/ha) accounts for nearly the total annual difference of 68 (g/ha)/yr. A similar comparison of loads from these basins during the spring melt (March-May), when hydrologic regimes and chemical quality of lake waters are most similar and the groundwater contribution relative to surface runoff in outflows is insignificant, indicates that the neutral basin accumulated 32 g/ha of lead, and the acid basin only 23 g/ha. Therefore, under similar hydrologic conditions, differences in lead accumulation in the acid and neutral basin are insignificant and differ by only 9 g/ha. From this analysis, differences between lead accumulation in the acid and neutral basins are hypothesized to be independent of pH and the degree to which acid precipitation is neutralized within the basins, but is rather a function of the hydrologic characteristics and groundwater regimes in these basins. Studies by Davis [24] on lead desorption from sediments in the acid and intermediate lakes further support the hypothesis that lead transport is independent of pH and indicate that lead desorption from lake sediment does not occur in water having a pH above 3.0, which is well below the ambient pH of these Adirondack Lakes.

Manganese and Zinc

Transport of manganese and zinc from the three watersheds correlates well with the degree of neutralization of acid precipitation. For manganese, the ratio of atmospheric inputs to basin outflows for the acid (0.38), intermediate (0.46), and neutral (0.74) lake basins indicates a net transport from all three basins (see Figure 6). The largest annual transport of manganese occurred in the acid basin (259 (g/ha)/yr) and was four times greater than that of the neutral basin (59 (g/ha)/yr).

Correlative with the relationship between manganese transport and lake pH, the ratios of annual input to outflow loads of zinc indicate a net accumulation in the neutral basin (1.4) and a net transport from the acid basin (0.69) and intermediate basin (0.74). The neutral basin accumulated zinc at a rate of 90 (g/ha)/yr whereas the acid basin lost 75 (g/ha)/yr. Thus, the annual difference in zinc transport between the neutral and acid basin was 165 (g/ha)/yr (see Figure 7).

Transport of both manganese and zinc was highest in the acid basin and lowest in the neutral basin. These findings closely parallel the expected increased solubility of manganese, and to a certain extent zinc, in waters of low pH [5, 6]. Although a similar correlation with pH was expected but not observed with iron, Hem [25] points out that as acid waters become neutralized, manganese stays in solution longer than iron. In the case of zinc, the solubility is higher than the observed levels, which indicates that

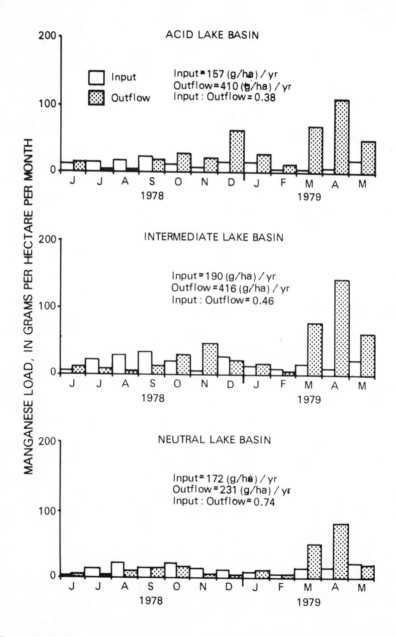

Figure 6. Monthly input and outflow loads of manganese from acid, intermediate and neutral basins. Unshaded bars indicate input; shaded bars indicate outflow.

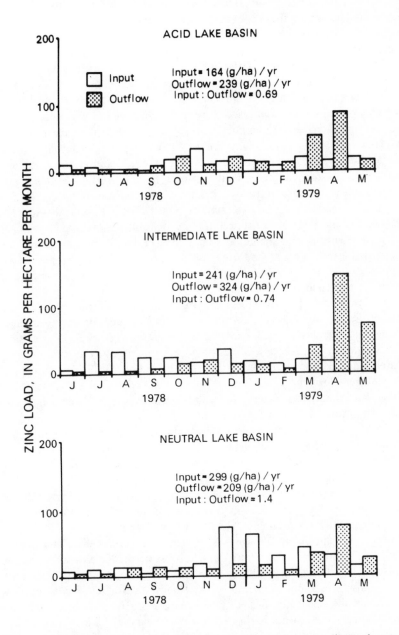

Figure 7. Monthly input and outflow loads of zinc from acid, intermediate and neutral basins. Unshaded bars indicate input; shaded bars indicate outflow.

other mechanisms such as ion exchange parallel predicted solubility controls. These findings suggest that the extent to which acid precipitation is neutralized within a watershed significantly affects the weathering and transport of manganese and zinc.

The relationship between pH and ion mobilization is further supported by manganese abundance in the sediments (Table 3). Sediments of the acid basin contained undetectable amounts of manganese, which would be expected if mobilization were caused by lower pH, whereas the sediments of the neutral watershed had an abundance of manganese, with a mean of 0.21 wt % of manganese oxide. However, this result may be fortuitous because the analytical procedure for the sediments of the acid and intermediate lake is less sensitive for manganese than that used for the neutral lake sediments.

The range in concentration of manganese in bulk precipitation samples from all three watersheds was relatively small (Table 8) and, unlike lead, showed no apparent seasonal trend. Zinc concentrations in bulk samples were somewhat more variable, as indicated in Table 9; no relationship between manganese or zinc and quantity of precipitation was observed. Mean annual zinc inputs of 235 (g/ha)/yr are comparable to input loads of 472 (g/ha)/yr measured at a rural Indiana site [26].

The histograms for manganese (Figure 6) and zinc (Figure 7) indicate a net accumulation of these elements in all three watersheds during summer (June–August); this is attributed to a combination of adsorption or organic complexing in the forest soils [22], bioaccumulation in the lake and (or) watershed [25], or accumulation in the lake sediments [2, 3]. During fall (September–November), outflow loads of these elements increased in all three watersheds; this increase is hypothesized to result from leaching and decomposition of leaf litter and other organic matter, which would remobilize previous supplies of bioaccumulated manganese and zinc [27]. Despite a lack of supporting data, the higher summer outflow loads from the neutral basin probably result from the higher contribution of manganese and zinc from groundwater in that basin than in the other basins.

A net transport of manganese from all three basins was observed during winter and spring (December–May) (Figure 6); the maximum outflow loads coincided with spring runoff (March–May) and accounted for 56, 67 and 68% of the annual outflow loads from the acid, intermediate and neutral basins (Table 6). Zinc transport during the same winter and spring period (December–May) differed slightly from the manganese pattern. Zinc transport during this 6-month period accounted for 80% of the annual outflow load from the acid basin, 86% of that from the intermediate basin, and 78% of that from the neutral basin (Table 6). The neutral basin received more than twice the atmospheric input of zinc than either of the other basins during this period; this resulted in a net accumulation of zinc in that basin.

Table 8. Summary of Manganese Data from Acid, Intermediate and Neutral Basins

	Streamflow				Precipitation		
	Acid	Intermediate	Lost Brook[a]	Neutral	Acid	Intermediate	Neutral
Mean concentration ± standard error (μg/L)	65 ± 2.7	55 ± 2.8	55 ± 5.3	39 ± 3.3	12 ± 2.6	21 ± 3.8	19 ± 3.3
Range in concentration (μg/L)	40–150	10–90	20–110	0–130	0–20	10–40	10–30
Number of samples	43	42	29	52	9	9	8
Total annual flux (outflow minus inflow) (g/ha)	410	416	405	231	151	190	172

[a]Lost Brook is inflow to neutral lake.

Table 9. Summary of Zinc Data from Acid, Intermediate and Neutral Basins

	Streamflow				Precipitation		
	Acid	Intermediate	Lost Brook[a]	Neutral	Acid	Intermediate	Neutral
Mean concentration ± standard error (μg/L)	44 ± 4.0	31 ± 5.3	32.7 ± 5.4	34 ± 4.5	14 ± 2.9	22 ± 3.5	16 ± 7.9
Range in concentration (μg/L)	10–110	0–220	10–160	0–180	0–30	10–50	0–60
Number of samples	43	42	29	50	9	9	8
Total annual flux (outflow minus inflow) (g/ha)	239	324	311	209	164	241	299

[a]Lost Brook is inflow to neutral lake.

Runoff during the spring (March–May) is generally sustained by the gradual melting of the snow pack, resulting in the release of highly acid waters to streams and lakes. Because the solubility of manganese and zinc increases under acid conditions, a greater weathering and transport of these constituents from all three watersheds was expected. During March-May 1979, 56, 67 and 68% of the total annual outflow load of manganese from the acid, intermediate and neutral lake basins occurred, and 62, 78 and 61% of the annual outflow load of zinc occurred (Table 6).

SUMMARY AND CONCLUSIONS

Iron transport from the three lake basins was independent of lake acidity or buffering capacity of the basins. Annual input loads of iron were 2332 (g/ha)/yr in the acid basin, 1441 (g/ha)/yr in the intermediate basin and 1346 in the neutral basin, as compared with outflow loads of 1065 (g/ha)/yr, 2110 (g/ha)/yr and 1155 (g/ha)/yr, respectively. No consistent seasonal trend in iron deposition was observed; comparison among basins was complicated by the large monthly variability in loading. The higher loads from the larger intermediate-pH basin may result, in part, from complexed iron that is accumulated in bogs and was transported from the basin during high flows.

Annual and monthly deposition of lead was similar in all three lake basins and averaged 183 (g/ha)/yr. Lead deposition was slightly higher during summer and fall than in winter and spring; on an annual basis, the acid basin accumulated 85% of its total lead input load, the intermediate basin 70% and the neutral basin 69%. Differences in lead retention among basins seem to be independent of mean annual lake pH and are hypothesized to be a function of the quantity and extent of groundwater recharge and seepage.

Manganese and zinc transport among basins seems to be directly related to the extent of neutralization of atmospheric precipitation. Mean input loads of manganese in each basin were similar and averaged 171 (g/ha)/yr; mean zinc input loads averaged 235 (g/ha)/yr and exhibited somewhat greater monthly variability than manganese. No seasonal trend in atmospheric deposition was observed for either element.

Outflow minus input loads of manganese from the acid lake basin was 259 (g/ha)/yr, from the intermediate basin was 226 (g/ha)/yr and from the neutral basin was 59 (g/ha)/yr. Outflow minus input loads of zinc from the acid and intermediate basins indicate a net transport of 75 (g/ha)/yr and 83 (g/ha)/yr, but in the neutral basin zinc accumulated at a rate of 90 (g/ha)/yr. The large differences in manganese and zinc transport among basins are hypothesized to reflect the degree of neutralization of acid precipitation by the unconsolidated glacial material.

Bioaccumulation of manganese and zinc seems to occur in each basin during summer, when biological activity is highest. These supplies are apparently remobilized and transported from the basin during the leaf-loading period in fall.

Iron, lead, manganese and zinc transport from all three watersheds was highest during the spring melt period; approximately 45% of the total iron and lead transport, and more than 65% of the manganese and zinc transport, coincided with spring melt. These metals combined with acid melt waters from the snowpack flow into the lakes, remaining on the surface of the stratified lakes, and discharging through the outlet. This flow path increases transport of metals from the basins during the spring melt period. The greater transport of manganese and zinc than of iron and lead during spring melt is attributed to the increased solubility of these metals in acid waters.

ACKNOWLEDGMENTS

This study is a contribution to the Integrated Lake Watershed Acidification Study (ILWAS) and was conducted in cooperation with the University of Virginia and the Electric Power Research Institute (RP-1109-5). Thanks are extended to William Kelly of Amherst College for providing the bedrock mapping, petrographic studies and microprobe analyses, and to the other participants in ILWAS for providing data and interpretation on related aspects of the lake watersheds.

REFERENCES

1. Beamish, R. J., and J. C. Van Loon. "Precipitation Loading of Acid and Heavy Metals to a Small Acid Lake Near Sudbury, Ontario," *J. Fish. Res. Bd. Can.* 3:649–658 (1977).

2. Henriksen, A., and R. F. Wright. "Concentrations of Heavy Metals in Small Norwegian Lakes," *Water Res.* 12(2):101–112 (1978).

3. Galloway, J. N., and G. E. Likens. "Atmospheric Enhancement of Metals in Deposition in Adirondack Lake Sediments," *Limnol. Oceanog.* 24(3):427–433 (1979).

4. Likens, G. E. "Acid Precipitation," *Chem. Eng. News* 54(48):29–43 (1976).

5. Garrels, R. M., and C. L. Christ. *Solutions, Minerals, and Equilibria* (New York, NY: Harper Row, 1964), p. 450.

6. Jenne, E. A. "Controls on Mn, Fe, Co, Ni, Cu and Zn Concentrations in Soils and Water–The Significant Role of Hydrous Mn and Fe Oxides," in *Trace Inorganics in Water*, Baker, R. A., ed. (Washington, DC: American Chemical Society, Advances in Chemistry Series No. 73, 1968), pp. 337–387.

7. Likens, G. E., et al. "Acid Rain," *Scientific Am.* 241(4):43–51 (1979).
8. Galloway, J. N., et al. "An Analysis of Lake Acidification Using Annual Budgets," in *Proceedings of an International Conference on the Ecological Impact of Acid Precipitation, Sandefjord, Norway, 1980* (Oslo, Norway: SNSF Project, 1980), pp. 254–255.
9. Cronan, C. S. University of Maine, Unpublished results (1979).
10. Newton, R. M. Smith College, Unpublished results (1980).
11. Johannes, A. H. Rennselaer Polytechnic Institute, Unpublished results (1979).
12. Hendrey, G. R. and L. Conway. Brookhaven National Laboratory, Unpublished results (1981).
13. Hendrey, G. R., J. N. Galloway and C. L. Schofield. "Temporal and Spatial Trends in the Chemistry of Acidified Lakes Under Ice Cover," in *Proceedings of an International Conference on the Ecological Impact of Acid Precipitation, Sandefjord, Norway, 1980* (Oslo, Norway: SNSF Project, 1980), pp. 266–267.
14. Likens, G. E., et al. *Biogeochemistry of a Forested Ecosystem* (New York, NY: Springer Verlag, 1977), p. 146.
15. Skougstad, M. W., et al. "Methods for Determination of Inorganic Substances in Water and Fluvial Sediments," U.S. Geological Survey Techniques of Water-Resources Investigations, Book 5, Chapter A–1 (Washington, DC: U.S. Government Printing Office, 1979).
16. Friedman, L. C., L. J. Schroeder and V. J. Janzer. "Programs to Assure the Quality of Water-Quality," data of the U.S. Geological Survey, *U.S. Geological Survey Circular* (in press).
17. "Water Resources Data for New York—Part I, 1978–1979." U.S. Geological Survey.
18. Lazarus, A. L., E. Lorange and J. P. Lodge. "Lead and Other Metal Ions in United States Precipitation," *Environ. Sci. Technol.* 4(1):55–58 (1970).
19. Pierson, D. H., et al. "Trace Elements in the Atmospheric Environment," *Nature* 241:252–256 (1973).
20. Husain, L., and P. J. Sampson. "Long-Range Transport of Trace Elements," *J. Geophys. Res.* 84(C3):1237–1240 (1979).
21. Siccama, T. G., and W. H. Smith. "Lead Accumulation in a Northern Hardwood Forest," *Environ. Sci. Technol.* 12(5):593–594 (1978).
22. Reiners, W. A., R. H. Marks and P. M. Vitousek. "Heavy Metals in Subalpine and Alpine Soils of New Hampshire," *Oikos* 26(3):264–275 (1975).
23. Wershaw, R. L. "Organic Chemistry of Lead in Natural Water Systems," in *Lead in the Environment*, U.S. Geological Survey Professional Paper 957, T. G. Lovering, Ed. (Washington, DC: U.S. Government Printing Office, 1976).
24. Davis. A. "Factors Controlling Lead Accumulation in the Sediments of Two Remote Adirondack Lakes," MS thesis, University of Virginia, Department of Environmental Sciences, Charlottesville, VA (1979), p. 81.
25. Hem, J. D. "Study and Interpretation of the Chemical Characteristics of Natural Water," U.S. Geological Survey Water-Supply Paper 1473 (Washington, DC: U.S. Government Printing Office, 1975).

26. Parker, G. R., W. W. McFee and J. M. Kelly. "Metal Distribution in Forested Ecosystems in Urban and Rural North-western Indiana," *J. Environ. Qual.* 7(3):337–342 (1978).
27. Slack, K. V., and H. R. Feltz. "Tree Leaf Control on Low Flow Water Quality in a Small Virginia Stream," *Environ. Sci. Technol.* 2(2): 126–131 (1968).

CHAPTER 5

DISTRIBUTION OF MERCURY AND FOURTEEN OTHER ELEMENTS IN REMOTE WATERSHEDS IN THE ADIRONDACK MOUNTAINS

G. Wolfgang Fuhs, Michael M. Reddy*
and Pravin P. Parekh

Division of Laboratories and Research
New York State Department of Health
Albany, New York 12201

INTRODUCTION

Acid precipitation has a number of undesirable effects. Two important health-related effects are: (1) an increase in lead and copper concentrations in certain drinking waters to a point reaching or slightly exceeding U.S. drinking water standard [1] and (2) the accumulation of mercury in fish [2, 3]. A currently favored hypothesis is that acidification favors the biological conversion of mercury into monomethyl mercury, the form more readily adsorbed by fish than the dimethyl form, to which mercury converts under less acidic conditions [4]. The present study is a geochemical reconnaissance of three large lake watersheds that are located adjacent to one another in the central Adirondacks and are all affected by acid rain (Figure 1).

DESCRIPTION OF STUDY AREA

The three lakes, from north to south, are Cranberry Lake, Low's Lake (formerly Bog River Flow) and Stillwater Reservoir (Figure 2). In Cranberry

*Present address: U.S. Geological Survey, Denver Federal Center, Denver, Colorado 80002.

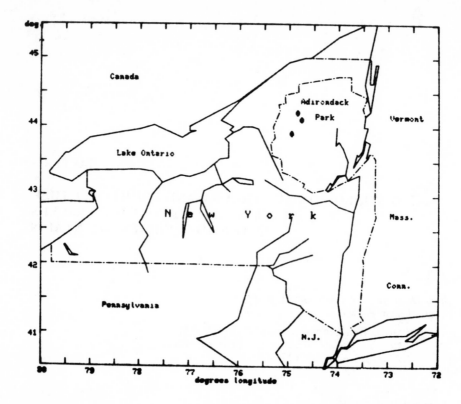

Figure 1. Adirondack study area park boundaries and location of study area in New York state. ● indicates location of study area.

Lake and Stillwater Reservoir the mercury concentrations in fish reach or exceed actionable levels [2, 5-7], but concentrations are low ($\leqslant 0.2$ mg/kg) in several trout taken in Low's Lake.

The lake basins are completely forested and very sparsely populated around the shores; the tributary watersheds are uninhabited. The only noteworthy events related to man's activities in recent history are a major forest fire in 1903 that devastated much of the Low's Lake watershed and parts of the Cranberry Lake watershed and some ongoing lumbering in the watershed of Low's Lake.

The basin is geologically Precambrian and consists of granite gneiss containing primarily feldspars, quartz and lesser amounts of hornblende pyroxenes and iron-titanium oxide minerals. In the southern part of the area this rock is interlayered with quartz syenitic gneisses, similar to the granite gneisses but with somewhat less quartz. These rocks underlie at least 75% of the Cranberry Lake and Stillwater Reservoir watersheds [3, 8]. In addition

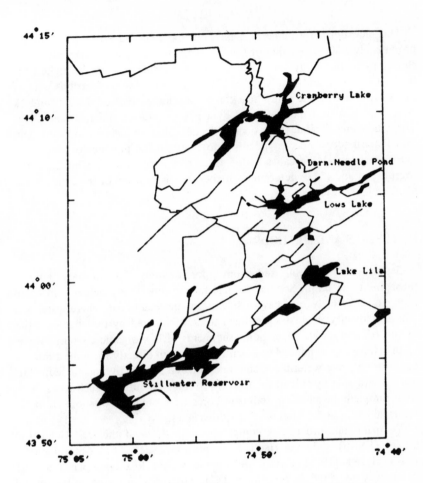

Figure 2. Detail of Adirondack study area: lakes and their tributaries.

there are several basin-like areas underlain by metamorphosed sedimentary and volcanic rocks. Several large faults and fractures influence drainage patterns. There is no evidence for geologically recent movement associated with these fractures. Glacial till and outwash lie in the lower elevations, most notably west and south of Low's Lake and northeast of Cranberry Lake [3].

The western Adirondacks are deeply wooded terrain with no volcanic or geothermal activity recent enough to explain the existence of point sources of mercury [3]. No specific data exist for mercury concentrations in Adirondack rocks or for comparable formations elsewhere in the world. Analyses for the more abundant elements were reported by Buddington and Leonard [8].

The elevation of Cranberry Lake and Stillwater Reservoir are both approximately 500 m and that of Low's Lake, approximately 550 m. Mountain elevations in the study area range from 700 to 800 m, with the highest between Cranberry and Low's Lakes. There are no greater elevations west of the area; therefore the study area can be considered part of the Adirondack west slope, which is most affected by acid precipitation and acid runoff [9]. The study area includes many small feeder lakes at higher elevations. Our geochemical survey was intended (1) to ascertain the presence or absence of local geochemical anomalies in the trace metal compositon of bedrock and overburden and (2) to determine whether the distribution of trace metals was affected by acid rain.

MATERIALS AND METHODS

Sediments were collected from approximately 50 tributary mouths throughout the basin. Sediments were taken immediately below each tributary mouth, usually in a bay or sidearm of the main lake, where water and sediment quality could be presumed to be dominated by the characteristics of the tributary watershed. Some sediments of low organic content (henceforth referred to as mineral sediments) were taken directly from the tributary bed near the lake waterline. Sometimes mineral and highly organic sediments were found side by side at or near the tributary mouth; in such cases, both were sampled. In addition, sediment cores were obtained offshore from the principal basins and subbasins of the three lakes.

To make our procedures compatible with those of the U.S. Geological Survey (USGS), only the fraction comprising particles of <2-mm diameter was analyzed [10, 11]. Dry weight and weight loss upon ignition were determined by standard procedures [12]. Total mercury, zinc, cadmium and lead were determined by concentrated HNO_3-H_2O_2 acid extraction and atomic absorption spectrophotometry (AAS) [13]. Lead, zinc and 10 other metals were also determined by X-ray fluorescence (XRF) analysis [14]. As expected, XRF analysis of lead and zinc gave slightly higher values than AAS, but otherwise the two sets of data were very highly correlated.

During a period of dry weather runoff in April 1979, pH values in the tributaries were monitored in a quasisynoptic fashion. All tributaries to each lake were surveyed in 1 or 2 days and all three lakes were covered in a 2-week period. Aluminum in the tributary waters was measured by AAS [12].

To examine the effects of acid leaching and artificial weathering of local gneiss bedrock, a fresh, unweathered specimen of representative leucogranitic gneiss from the Cranberry Lake watershed was ground and sieved. Sieved fractions were washed with distilled water and air dried. The procedure outlined by Berner et al. [15] was followed. Briefly, samples were reacted with 5% HF plus 12% HCl solution for 1-20 hr. Teflon®* labware, reagent-grade chemicals and distilled water were used throughout the investigation. At specific intervals during each experiment the supernatant was removed from the reaction beaker and evaporated to dryness. The residue was taken up in 100 mL of dilute HCl and analyzed for metal content by AAS. Total metal content was determined after complete dissolution in concentrated HF and HNO_3. Unreacted rock was ground and then characterized by polarized light microscopy with dispersion staining. Scanning electron microscopy was used to examine the surface of both reacted and unreacted mineral grains.

RESULTS

Sediment Chemistry

Concentrations of the toxic elements zinc, cadmium, mercury and lead were strongly correlated with the organic content of the sediments (Figure 3). The element ratios varied considerably from sample to sample, but generally the slopes of the regression lines of metal versus organic content were almost identical. We consider mineral sediments as those with an organic content below 7% and organic-rich sediments as those with 7% or higher weight loss on ignition. Organic-rich sediments from shallow waters (samples collected from the tributary mouths or their immediate vicinity) are discussed separately from deepwater sediments, which are all organic in nature and were collected from deep water, either with an Ekman dredge or as the top centimeter of a sediment core. The metal contents of organic-rich sediments are expressed on an ash basis.

For most of the metals, variations in the concentrations reached over two orders of magnitude. Comparison of the results from the small watersheds by geographic area showed no significant local anomalies, with the possible exception of elevated mercury concentrations in sediments from the area where the watersheds of Cranberry Lake and Low's Lake meet.

The chemical results from all watersheds (Figure 4) show elevated concentrations of several metals (iron, zinc, lead, cadmium and mercury) in the organic shallow sediments, even after correction for organic content. Even higher concentrations are found in the offshore sediments. Unlike the local differences, many of these differences are significant. Figure 4 also shows our analysis of the unweathered bedrock sample and the values published for

---

*Registered trademark of E. I. du Pont de Nemours & Company, Inc., Wilmington, Delaware.

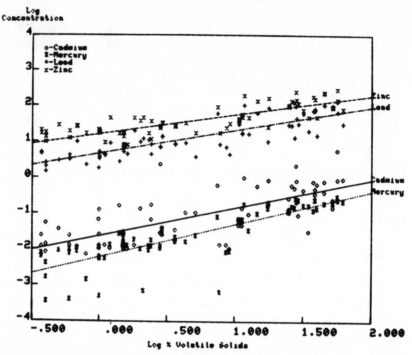

Figure 3. Least-squares regression plot of cadmium, lead, mercury and zinc sediment concentrations as a function of sediment volatile solids content. Regression equations and correlation coefficients are: Cd, $Y = 0.79X - 1.66$ ($r = 0.73$); Pb, $Y = 0.67X + 0.65$ ($r = 0.83$); Hg, $Y = 0.96X - 2.29$ ($r = 0.85$); Zn, $Y = 0.49X + 1.27$ ($r = 0.80$); where Y is the log concentration of the respective element and X is the log % volatile solids content of the sediment.

USGS granite standard G1 [16] which is gray and contains less iron and manganese than the typically reddish-brown Adirondack rock.

Sediment-Cores

Analysis of three sediment cores from each lake shows enrichment (an increase by one order of magnitude or more) of lead in the sediment surface, relative to the deeper layers (Table 1). Some apparent anomalies in the Stillwater Reservoir zinc concentrations actually are anomalies in total metal content, as they appear in both AAS and XRF analyses. Cadmium concentrations at the surface are up to fivefold higher, mercury and zinc are only two- to threefold higher. In Low's Lake a fivefold increase of copper in the sediment surface is notable in connection with the leaching experiments discussed below. Iron and manganese are slightly increased; the remaining elements are uniformly distributed with depth. Most sediments are highly organic. The above statements apply to both the dry weight and ash basis results.

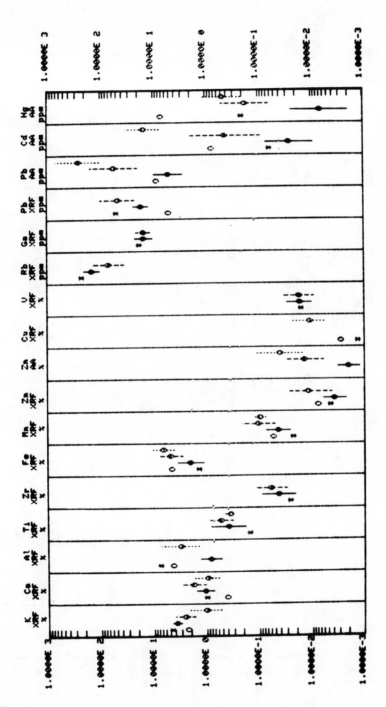

Figure 4. Elemental abundances in central Adirondack sediments. * = granite standard; —— = mineral sediments; – – – = organic-rich sediments in shallow water; and … = top layer of lake cores. All organic-rich sediments are on an ash basis; 0 = Adirondack leucogranitic gneiss; error bars correspond to one standard deviation.

Table 1. Selected Trace Metals in Sediment Cores from Central Adirondack Lakes

Depth (cm)	Cranberry Lake[a]				Low's Lake[b]		Stillwater Reservoir[c]		
	V	D	J	L	G	M	H	R	D
Mercury (AAS) ($\mu g/g$)									
0-1	0.36	0.28	0.73	0.81	0.74	0.86	0.08	0.48	0.95
1-5		0.31	0.40	1.35	0.74	0.33	0.08	0.31	0.33
5-10		0.20	0.29	0.67	0.25	0.43	0.09	0.11	0.42
10-15	0.31	0.22	0.32	0.48	0.31	0.69	0.18	0.11	0.54
15-20		0.12	0.30	0.28	0.28	0.43	<0.08	<0.08	
20-25	0.21	0.27	0.42	0.36	0.20	0.19		<0.08	
25-30	0.22	0.28	0.11	0.24	0.34	0.18			
30-35		0.19	0.15			0.16			
35-40		0.22				0.14			
40-45		0.12							
Zinc (AAS) ($\mu g/g$)									
0-1	359	710	531	432	691	552	37	482	762
1-5		175	398	357	1033	312	45	454	426
5-10		245	208	277	431	388	38	212	462
10-15	208	340	295	285	343	489	38	110	254
15-20		311	231	275	338	352	31	40	
20-25	178	326	134	249	199	324		30	
25-30	189	489	183	—	172	73			

Depth								
30–35	459	170			35			
35–40	220				32			
40–45	245							

Lead (AAS) (µg/g)

Depth								
0–1	290	379	541	294	517	41	108	635
1–5	63	199	297	443	125	21	57	263
5–10	111	156	225	123	171	20	56	168
10–15	100	88	187	114	122	20	44	154
15–20	156	83	196	85	78	20	21	
20–25	102	28	142	28	94		21	
25–30	122	27	137	<28	46			
30–35	83	26			20			
35–40	60				<20			
40–45	61							

{ 134 (0–1), 61 (1–5), 18 (20–25), 38 (25–30) }

Cadmium (AAS) (µg/g)

Depth								
0–1	145	10	41	10	34	4	11	16
1–5	4.2	6	8	22	8	<4	6	7
5–10	4.5	5	7.5	6	8.5	<4	6	8
10–15	6	6	7.5	6	8	<4	4	<15

{ 3.2 (0–1), 1.2 (5–10) }

Table 1, continued

Depth (cm)	Cranberry Lake[a]		Low's Lake[b]				Stillwater Reservoir[c]		
	V	D	J	L	G	M	H	R	D
15-20		6	6	8	6	8	<4	<4	57
20-25	0.78	6	6	<7	<6	9		<4	100
25-30	0.86	6	5	7	<6	9			44
30-35		4	5			<3.5			—
35-40		4				<3.5			—
40-45		4							—
Copper (XRF) (μg/L)									
0-1		117	62	84	241	398	—	69	
1-5		—	—	—	—	44	—	—	
5-10		33	58	133	68	—	—	—	
10-15		31	31	64	59	57	—	—	
15-20		47	—	69	49	—	—	—	
20-25		22	42	71	48	—	—	—	
25-30		40	16	39	—	64	—	6	
30-35		45	17						
35-40		45							
40-45		19							
Volatile Solids (%)									
0-1	27.7	31	21	26	32	42	1.5	17	37
1-5		4.08	29.6	52.9	72.9	51.9	2.7	29.5	39.1

Depth (cm)									
5–10	28.0	10.3	4.08	46.6	35	53.1	10.2	29.1	52.4
10–15	27.1	0.39	32.2	46.9	30	50.9	1.1	8.8	74.1
15–20		35.8	27.5	49.1	29	48.8	2.0	4.3	
20–25	25.8	2.0	28.5	43.8	29.7	57.4		5.9	
25–30		34.6	26.9	41.4	30.2	56.4			
30–35		4.12	22.4			48.4			
35–40		2.19				42.9			
40–45		6.10							

[a] Cranberry Lake: V = near Cranberry Lake Village (core cut in 7.5-cm segments), D = Dead Creek flow, J = off Joe Indian Island.

[b] Low's Lake: L = Low's Lake (= Bog River flow) main channel off Graves Mountain, G = Grass (or Grassy) Pond, deepest part (NE corner), M = Mud Lake portion of Low's Lake.

[c] Stillwater Reservoir: H = between tributaries from Mud and Hidden Ponds, R = west of Little Rapids, D = near Dam.

The sedimentation rates were determined by the ^{210}Pb method [17] in one core from the lower part of Stillwater Reservoir (43°54'17"N, 74°58'45"W) and the central section of Low's Lake (44°05'16"N, 74°44'49"W). The results were 0.09 and 0.04 g/cm^2-yr, respectively, but in the case of Stillwater Reservoir the supported ^{210}Pb (i.e., the portion derived from local ^{226}Ra) is suspected to be a substantial fraction of the total so that the actual sedimentation rate was probably close to that of Low's Lake. These relatively low sedimentation rates are consistent with the character of the two watersheds with their dense forest cover and only few subwatersheds yielding significant amounts of suspended material.

Watershed pH, Dissolved Aluminum

Results of the synoptic survey of lake tributary pH values yield information about acid inputs to the study area. Figure 5 shows the slopes of the tributary beds in the Cranberry Lake watershed. A quite similar graph can be obtained from the tributaries of Low's Lake, which is located immediately south of Cranberry Lake and shares with it several mountains with steep slopes and outcroppings of bare rock. A graph of tributary pH against the

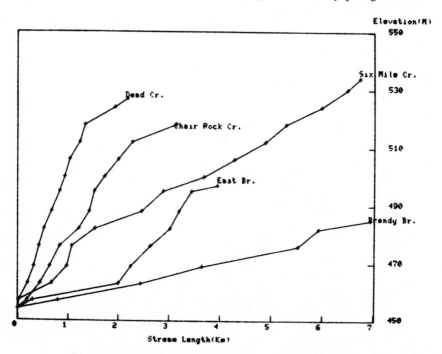

Figure 5. Slopes of stream beds tributary to Cranberry Lake.

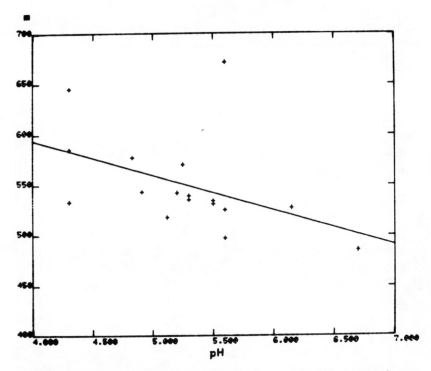

Figure 6. pH of water in Cranberry Lake tributaries plotted as a function of source elevation (r = –0.457, 15 d.f.).

elevation of the source (Figure 6) shows an inverse correlation of borderline significance. The lowest pH values (4.2) were observed on the slopes of the higher elevations, where soil cover is minimal or nonexistent. The highest pH values (6.7) were found in tributaries that have gentle slopes and flow through valleys with extensive glacial deposits. The higher elevations in the Adirondacks are in frequent contact with highly acid clouds that may contribute to the low pH observed in runoff from the mountain slopes. However, extensive contact with soil formations, particularly in the form of glacial overburden, seems to be the single most important factor that determines the acidity of watershed runoff.

Dissolved aluminum ion has been proposed by Schofield [18] and others as a major toxic product of acid rain reaction with surface minerals. To examine this possibility, soluble aluminum in watersheds was plotted against watershed acidity (Figure 7). The observed correlation is in agreement with the results of Schofield and others.

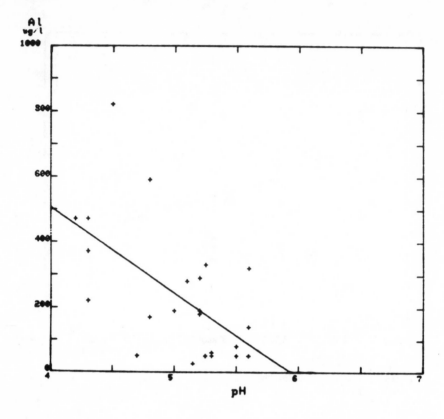

Figure 7.   Dissolved aluminum concentration in Cranberry Lake tributaries versus source elevation (r = –0.583, 21 d.f.).

Simulated Weathering–Acid Leaching Experiments

Treatment of unweathered leucogranitic gneiss from Cranberry Lake (Figure 8) with HF and HCl produces a gradual dissolution of the elements associated with the rock matrix, such as aluminum, calcium and iron. By contrast, well-defined fractions of lead, copper, mercury, zinc and cadmium are removed within the first hour of the treatment, with very little subsequent leaching. The results in Figure 8 are based on the total elemental analysis of the same sample which is included in Figure 4. Noteworthy is the substantial removal of lead and copper from this 0–2 mm fraction. These observations may reflect the preferential dissolution of a trace-metal-enriched reactive phase in the crushed rock.

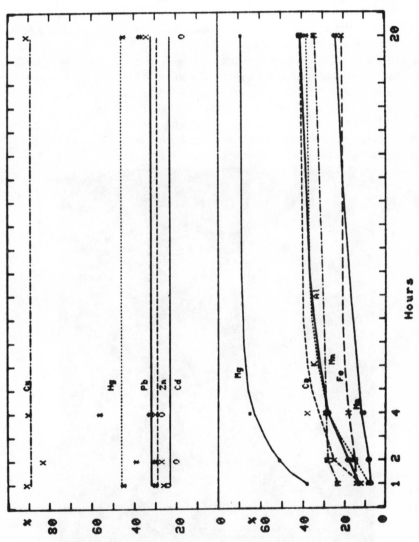

Figure 8. Results of simulated weathering and acid-leaching experiments with Adirondack leucogranitic gneiss.

Microscopic Examination of Bedrock

The granite gneiss, after crushing and sieving, yielded two identifiable grain types. Quartz was the major constituent. Biotite, an iron-rich micaceous aluminosilicate, and hornblende also were present in significant quantities. Quartz, biotite and hornblende were positively identified by polarizing light microscopy with dispersion staining.

In the scanning electron microscope mineral grains of the crushed granitic gneiss exhibit transgranular fragmentation which reflects the fracture modes of the constituent minerals: angular and conchoidal for quartz, and cleavage for the feldspar and hornblende (Figures 9 and 10). Total analysis of the ground sample (Figure 4) indicates significant quantities of feldspars (aluminum silicate with sodium, potassium or calcium), in addition to the ferromagnesian minerals hornblende and biotite.

Figure 9. Low-magnification scanning electron micrograph of crushed Cranberry Lake leucogranitic gneiss. Scale marker 100 μm.

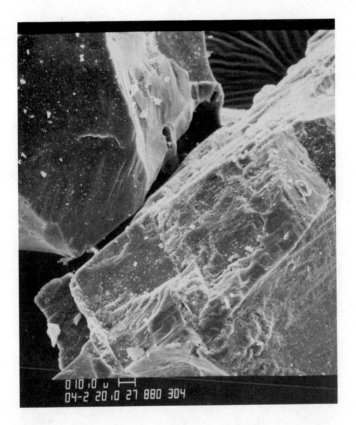

Figure 10. High-magnification scanning electron micrograph of center of Figure 9. Grain at upper left is quartz. Scale marker 10 μm.

After dissolution in the acid mixture, the mineral grains exhibit distinct surface morphology after 2-hr dissolution. (Compare Figure 10 top, un-reacted, with Figure 13.) Feldspars and hornblende, on the other hand, are readily attacked by the mixed acid and show the surface etching and devel-opment of etch pits characteristic of dissolution. (Compare Figure 10 (un-reacted) with Figures 11 and 12 (1 hr) and Figure 14 (2 hr).)

These micrographs also show that the major components of the Cranberry Lake watershed bedrock undergo dissolution at different rates. The com-ponents with potentially the highest trace metal content (hornblende and biotite) are most reactive and dissolve most rapidly. This observation is con-sistent with the results of the leaching experiments. Iron and aluminum arise from dissolution of the rock matrix of the feldspars and possibly also of the ferromagnesian minerals.

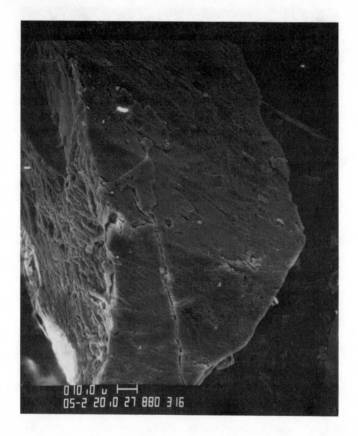

Figure 11. Etch pits in constituents of leucogranitic gneiss after 1-hr exposure to HF–HCl solution. Scale marker 10 μm.

DISCUSSION

Our geochemical survey of sediments from the 50 subwatersheds of three central Adirondack lakes revealed no major anomaly with regard to the elemental composition of the bedrock. However, the patterns of leaching and accumulation indicate that mechanisms of metal transport may be directly affected both by atmospheric transport of metals and by acidity in precipitation.

The two primary mechanisms which may affect trace metal concentrations in Adirondack streams and sediments are: (1) atmospheric deposition [19,20] on land and water, with subsequent leaching and transport of this atmospheric component from land to water, and (2) weathering and leaching from bed-

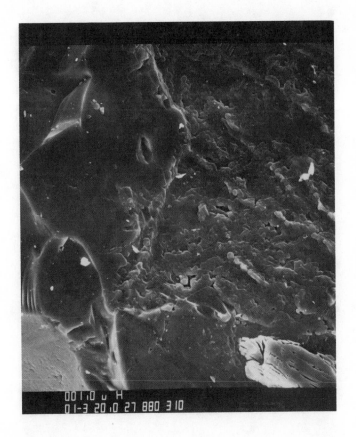

Figure 12. Same as Figure 11, high magnification. Scale marker 1.0 μm.

rock and overburden, followed by transport to receiving waters and sedimentation.

To discuss the possible role of these mechanisms, we can examine the distributions of the several elements shown in Figure 8 and the enrichment factors with respect to manganese shown in Table 2. The enrichment factors are calculated from the concentrations (weights on ash basis) as follows:

$$\frac{[M] \text{ sample}/[Mn] \text{ sample}}{[M] \text{ std.}/[Mn] \text{ std.}}$$

where M is the metal of interest and std. is the standard material as specified in Table 2. Manganese was chosen as the reference element because data were available for most samples, because it is thought to be of soil origin [21],

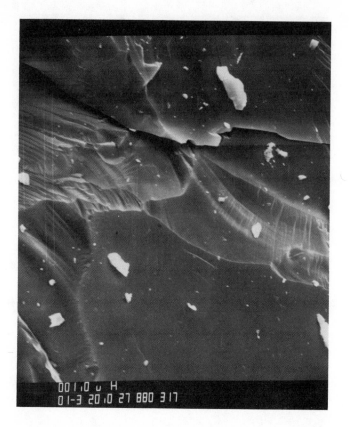

Figure 13. Quartz grain in crushed leucogranitic gneiss. No signs of attack after 2 hr of exposure to HF–HCl solution (compare with Figure 11, top). Scale marker 1.0 μm.

and because in the aerobic surface sediments of these oliogotrophic lakes we do not expect major errors from the solubilization of manganese under reducing conditions. Moreover, it is considered that iron and manganese oxides exist in a number of high-surface-area, high-reactivity forms in the environment. These oxides are excellent scavengers for trace metals [22]. Thus, normalization of trace metal concentrations to the concentration of manganese in the sediment seems to be appropriate. Factors over 10 are commonly accepted as indications of enrichment or depletion.

Although we had hoped to obtain additional information by relating watershed acidity from our synoptic survey to trace-element concentrations in mineral sediments, no statistical correlation was found. Only the occurrence of elevated concentrations of aluminum in the water can be related to acid leaching of bedrock and overburden. Our reference point for aluminum

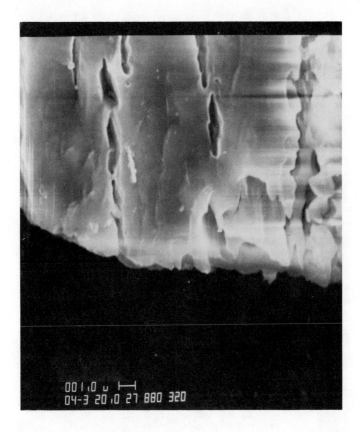

Figure 14. Severe surface erosion and etch pits in constituent of leucogranitic gneiss after 2-hr exposure to HF–HCl solution. Scale marker 1.0 µm.

in Figure 4 is conclusive in this respect, although the aluminum analysis was performed by HNO_3-H_2O_2 extraction and therefore does not represent total metal content. The leaching pattern (Figure 8), in combination with the great abundance of this element, also supports our conclusion.

Iron, manganese and zinc distributions resemble that of aluminum. The data in Figure 4 are representative for these elements in that both standards and sediments had been analyzed for total metal content by XRF. Since zinc is enriched in airborne particulates (Table 2), an atmospheric contribution of this element is a possibility. All three elements are more abundant in the organic-rich than in the mineral sediments.

Copper is removed from the bedrock rapidly and almost completely. Although atmospheric deposition of copper may be a factor, the enrich-

Table 2. Enrichment Factors Based on Manganese Abundance

Element	Mineral Sediments (Tributary) to Granite Standard	Mineral Sediments (Tributary) to Unweathered Rock	Organic-Rich Sediments (Tributary) to Mineral Sediments (Tributary)	Organic-Rich Sediments (Lake) to Mineral Sediments (Tributary)	Air Particulates[a] to Mineral Sediments (Tributary)
Aluminum	0.06	0.24		1.7	
Silicon	0.5			0.45	
Potassium	0.4	2.0	0.3	0.13	
Rubidium	0.33		0.3	0.2	
Calcium	0.55	3.2	0.7	0.4	
Strontium	0.37		0.6	0.4	
Titanium	1.35		0.6	0.4	0.17
Vanadium	5.6		0.6		0.85
Iron	0.2	0.55	1.0	1.5	0.63
Cadmium	0.3	0.2	4.1	160	400
Mercury	0.02	0.012	11.3	32	
Lead	0.19	3.2	1.3	3.4	63
Zinc	0.43	0.61	1.6	4.9	16

[a]Composition of air particulates over Lake Michigan [20]. All organic sediments on an ash basis.

ment of this element in the sediments in two parts of Low's Lake—Grass Pond and Mud Lake (Table 1)—is of interest because these parts receive the runoff with the lowest pH in the study area. Copper is prominently enriched in potable waters from shallow wells immediately south of the study area, reaching or exceeding the U.S. drinking water standards of 1 mg/L.

The detailed mechanism of trace metal-sediment interactions in acid waters remains an area of active research. Both free metal ions and complexes can exist in solution under ambient conditions. Hydrolytic reactions may cause metals to adsorb on particulate material such as hydrous metal oxides, clays and organic colloids. In addition, the presence of organic acids can stabilize lower oxidation states of some metal ions in solution. Precipitation and dissolution of hydrous metal oxides and minerals may be strongly influenced also by the presence of humic materials [23].

Cadmium is leached from bedrock, and the atmospheric component also may be significant (Table 2). Since the analysis of the sediments involves HNO_3-H_2O_2 extraction of cadmium in the mineral sediments and total digestion of the standard, the difference between the two in Figure 4 may be exaggerated. The enrichment in the organic-rich sediments is, however, significant. Cadmium concentrations in Adirondack lakes are in the order of 0.2-5 μg/g [9].

The mercury content of the unweathered Adirondack rock was significantly higher than that of the granite standard (7 vs 0.2 μg/g). The fine material sediments are depleted in mercury with respect to both these standards. This and the acid-leaching experiments indicate that mercury is readily removed from the parent rock. It subsequently accumulates preferentially in the organic sediments. The highest concentrations (0.5-0.6 μg/g) occur in Grass Pond and other western parts of Low's Lake. The atmospheric inputs are not known, but mercury is emitted during coal burning, is transported in both particulate and vapor form [24], and may temporarily accumulate in living matter.

Lead follows the same pattern. The low lead content of our Adirondack leucogranitic gneiss sample (6 μg/g) is surprising (Figure 4). Sixty percent of the lead was removed by acid leaching. The enrichment with lead of the organic and particularly the offshore sediments is apparent (Figure 4). For lead, there is certainly a significant atmospheric source (Table 2). This may explain the difference in total lead content by XRF between the unweathered rock and the fine mineral tributary sediments. The accumulation in offshore sediments can be related to direct atmospheric inputs into these parts, where sedimentation should be less rapid than in the inshore areas. It is also possible that lead and other metals are preferentially enriched in the finest particulates, which are carried farthest from shore. Parts of Low's Lake which show elevated heavy metal concentrations are located in areas of the most acidic runoff, but there is also a direct relationship between this acidity and the absence of soil cover on the steep, rocky mountain slopes, with the likelihood of low erosion and sedimentation rates. If airborne metals are deposited directly on the water, their accumulation in the offshore sediments will be greatest in areas of low sedimentation rates for the bulk minerals. The effects of selective heavy metal leaching and atmospheric deposition will reinforce each other.

This same combination of factors may explain isolated elevated concentrations of lead (0.1-0.2 mg/L) in local drinking waters [1] from wells and shallow springs. The concentrations of lead in lakewater are much lower [9].

Results of several recent studies support the suggestion that lead accumulation in surficial lake sediment, in part, is a manifestation of atmospheric deposition. Sherahata et al. [25] show that excess lead in surface sediment of a California mountain pond is of eolian anthropogenic origin and that the surface enrichment is irrefutably linked with industrial sources. These authors suggest that many such remote North American ecosystems are highly polluted by industrial lead aerosols. Lead accumulation in remote eastern ecosystems has also been reported. For example Siccama et al. [26] have shown that the forest floor in white pine stands of central Massachusetts experienced a net gain in lead content over the past 16 years. This accumulation is attributed to increasing lead inputs via atmospheric deposition. An extensive review of much of the recent literature dealing with lead in sediments has indicated to Nriagu [27] that "It is now generally accepted that the atmospheric input is the dominant source (60-80 percent) of lead in recent sediments".

ACKNOWLEDGMENTS

Most shores of Low's Lake are privately owned, and the lake is not easily accessible to the public. The authors express their sincere thanks to those owners who greatly facilitated access to the lake and to several tributary watersheds and provided the fish specimens for analysis.

We thank Dr. Philip Whitney of the New York State Geological Survey for valuable advice in the identification of Adirondack rock samples. The chemical extractions and AA analyses were performed under the supervision of Messrs. Rolf A. Olsen and Robert Weinbloom. Dr. Martin Wahlen of the Radiological Science Institute of this Division carried out the ^{210}Pb analyses of two sediment cores.

REFERENCES

1. Fuhs, G. W. "A Contribution to the Assessment of Health Effects of Acid Precipitation," in *Proceedings of the Action Seminar on Acid Precipitation, 1-3 November 1979, Toronto,* ASAP Organizing Committee, pp. 113-116.
2. Joint Advisory from the Departments of Health, Agriculture and Markets, and Environmental Conservation, New York State Department of Health (May 20, 1971).
3. Ad hoc Committee Appointed by the Commissioner, New York State Department of Health, "The Health Implications of Methyl Mercury in Adirondack Lakes," New York State Department of Health (February 28, 1978).
4. Fagerstrom T. and A. Jernelov. "Some Aspects of the Quantitative Ecology of Mercury." *Water Res.* 6:1193-1202 (1972).
5. U.S. Food and Drug Administration, Press Release 71-25 (May 6, 1971).
6. Armstrong, R. W., and R. J. Sloan. "Trends in Levels of Several Known Chemical Contaminants in Fish from New York State Waters," Technical Report 80-2, New York State Department of Environmental Conservation (June 1980), 77 pp.
7. "Environmental Health Criteria 1 (Mercury)," World Health Organization, (Geneva: 1976).
8. Buddington, A. F., and B. F. Leonard. "Regional Geology of the St. Lawrence County Magnetite District, Northwest Adirondacks, N.Y.," U.S. Geological Survey Professional Paper 376 (1962).
9. Wood, L. W., "Limnology of Remote Lakes in the Adirondack Region of New York State with Emphasis on Acidification Problems," New York State Department of Health, Environmental Health Report No. 4 (1978).
10. Skougstad, M. W., et al., Eds. "Techniques of Water Resources Investigations of the U.S. Geological Survey," in: *Methods for Determination of Inorganic Substances in Water and Fluvial Sediments,* U.S. Geological Survey, (1979), p. 25.
11. Krishnamurty, K. V., and M. M. Reddy. "The Chemical Analysis of Water and Sediments in the Genesee River Watershed Study," Environmental Health Center, Division of Laboratories and Research, New York State Department of Health (December 1975).
12. *Standard Methods for the Examination of Water and Wastewater,* 13th ed. (American Public Health Association, 1971), pp. 535-538.
13. Krishnamurty, K. V., E. Shpirt and M. M. Reddy, "Trace Metal Extraction of Soils and Sediments by Nitric Acid-Hydrogen Peroxide," *Atomic Adsorp. Newslett.* 15:68-70 (1976).
14. Parekh, P. P. "Energy-Dispersive X-Ray Fluorescence Analysis of Organic-Rich Soils and Sediments," *Radiochem. Radioanalyt. Lett.* (in press).
15. Berner, R. A., et al. "Dissolution of Pyroxenes and Amphiboles during Weathering," *Science* 207:1205-1206 (1980).
16. Mason, B. *Principles of Geochemistry,* 3rd ed. (New York: Wiley, 1966).

17. Wahlen, M., and R. C. Thompson. "Pollution Records from Sediments of Three Lakes in New York State," *Geochim. Cosmochim. Acta,* 44:333–339 (1980).

18. Schofield, C. L. "Acid Precipitation: Our Understanding of the Ecological Effects" in *Proceedings of the Conference on Emerging Environmental Problems: Acid Precipitation,* U.S. Environmental Protection Agency. Region II EPA–902/9–75–001 (1975), pp. 76–87.

19. Galloway, J., and G. E. Likens. "Atmospheric Enhancement of Metal Deposition in Adirondack Lake Sediments," *Limnol. Oceanog.* 24: 427–433 (1979).

20. Gatz, D. F. "Pollutant Aerosol Deposition into Southern Lake Michigan," *Water Air Soil Poll.* 5:239–251 (1975).

21. Andren, A. D., and S. E. Lindberg. "Atmospheric Input and Origin of Selected Trace Elements in Walker Branch Watershed," *Water Air Soil Poll.* 8:199–215 (1977).

22. Jenne, E. A. "Trace Inorganics in Water," *Adv. Chem.* 73:337–387 (1968).

23. Jackson, K. S., I. R. Jonasson and G. B. Skipper. "The Nature of Metals-Sediment-Water Interactions in Freshwater Bodies, with Emphasis on the Role of Organic Matter," *Earth-Sci. Rev.* 14:97–146 (1978).

24. Johnson, D. L., and R. S. Braman. "Distribution of Atmospheric Mercury Species Near Ground," *Environ. Sci. Technol.* 8:1003–1009 (1974).

25. Shirahata, H., et al. "Chronological Variations in Concentrations and Isotopic Composition of Anthropogenic Atmospheric Lead in Sediments of a Remote Subalpine Pond," *Geochim. Cosmochim. Acta* 44: 149–162 (1980).

26. Siccama, T. G., W. H. Smith and D. L. Mader. "Changes in Lead, Zinc, Copper, Dry Weight and Organic Content of the Forest Floor of White Pine Stands in Central Massachusetts over 16 Years," *Environ. Sci. Technol.* 14:54–56 (1980).

27. Nriagu, J. O. "Lead in Soils, Sediments and Major Rock Types," in *Biogeochemistry of Lead,* J. O. Nriagu, Ed. (Amsterdam: Elsevier/ North Holland Biomedical Press, 1978), pp. 15–72.

CHAPTER 6

PRECIPITATION ANALYSIS IN CENTRAL NEW MEXICO

Carl J. Popp, Claire M. Jensen and Donald K. Brandvold

Department of Chemistry
New Mexico Institute of Mining and Technology
Socorro, New Mexico 87801

Lynn A. Brandvold

New Mexico Bureau of Mines and Mineral Resources
Socorro, New Mexico 87801

INTRODUCTION

The presence of acidic precipitation (pH < 5.6) has been well established in many sections of the United States, with most of the research efforts centered in the northeastern portion of the country [1-3]. However, a large area in the intermountain region of the United States, including the high deserts and mountains of the Southwest, has received little attention, and there currently are no published data from a long-term study. Analysis of bulk precipitation in the mountains of northern Colorado by Lewis and Grant [4] indicated a steady drop in pH values over a recent three-year period, with the acidity attributed to nitric acid.

The southwestern United States offers considerable climate contrast with life zones ranging from Lower Sonoran Desert to Hudsonian (Alpine), often within a few miles of each other. Air quality is generally excellent, with vistas of 100 miles quite common. The region is characterized by calcareous soils and low population density. These areas are well known for their mineral wealth, including large coal deposits, and development of coal-burning power facilities is imminent, with increased demand for electricity. This region of the country is currently experiencing growth in coal-related energy activities [5]. The Four Corners area of New Mexico, in which two coal-fired power plants of >2000 MW capacity are currently operating, is responsible for >50 ktons of SO_2 emissions per year [1]. This is considerably greater than the sum of established SO_2 emissions combined from the adjoining states of Arizona, New Mexico, Utah and Colorado.

The study area for which these data have been collected is located in south-central New Mexico (see Figure 1). One collection site in the city of

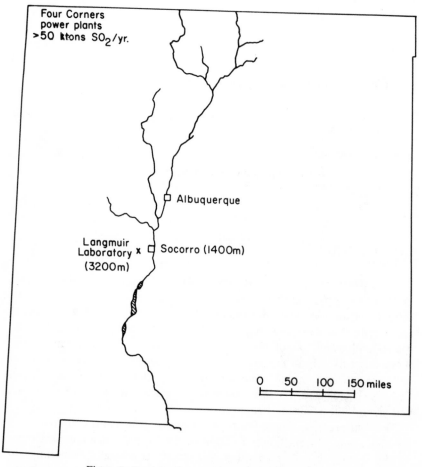

Figure 1. Precipitation sampling areas—New Mexico.

Socorro (pop. 7000) is located along the Rio Grande Valley at an altitude of 1400 m, with an average annual precipitation of 20 cm. The climate is characterized as Upper Sonoran life zone. The second sampling site is located directly west of Socorro, where the New Mexico Institute of Mining and Technology maintains the Irving Langmuir Laboratory for Atmospheric Research. The laboratory is located at the 3240-m level on a mountain ridge that is 27 km air distance from Socorro. Mean precipitation at the laboratory is 48 cm, with more than one-half (28 cm) coming in the summer months of June, July and August. The laboratory is in a spruce-fir life zone.

The purpose of this research was to examine precipitation chemistry in the central New Mexico portion of the southwestern United States to determine how widespread acid rain effects may be in this area and whether there are any unusual effects characteristic of the region. Sample sites were chosen in proximity to each other, but at very different altitudes, to study the effects on precipitation chemistry when rain falls through more of the earth's atmosphere. This can be important when blowing dust is present, as is often the case in these arid regions.

EXPERIMENTAL

Collection Device and Sample Collection

Because of the small amount of annual rainfall in the valley (20 cm) and during the summer collection season at Langmuir Laboratory (28 cm), it was necessary to construct large funnels to optimize sample volumes. The funnels were constructed completely of plastic, with the main body of 1/16-in. polyethylene sheeting cut into a circle from which a wedge was removed. The edges of the wedge cut were overlapped 3 cm and fastened with nylon screws. The final diameter of the funnel was 1 m, which allowed 2 L of water to be collected from a 2.5-mm rainfall. The funnel was covered with polyethylene sheeting between precipitation events to omit dry deposition and was uncovered and rinsed with deionized water immediately before an impending precipitation event. Samples were collected in 2-L acid-washed polyethylene bottles.

Sample Treatment and Analysis Procedures

Samples were brought to the laboratory immediately in the case of the valley samples or within a few hours if collected at the mountain laboratory. The pH was taken immediately using an Orion 601 pH meter calibrated with two buffers, and a portion of the sample was titrated for carbonate species with either sulfuric acid if pH > 5 or sodium hydroxide if pH < 5. The sample was then filtered through a 0.2-μ Millipore filter and a portion was acidified to pH 2-2.5 with redistilled HNO_3 for atomic absorption (AA) analysis. The remainder of the sample was stored at 4°C for major anion analysis.

The metal ions sodium, potassium, magnesium and calcium were analyzed by flame AA using a Perkin-Elmer 303 spectrophotometer. Most of the trace metals were analyzed using an HGA 2000 graphite furnace with a Perkin-Elmer 403 AA; selenium and arsenic were determined with a Perkin-Elmer EDL AA system. Procedures followed those outlined in the U.S. Environmental Protection Agency (EPA) manual EPA-600/14-79-020 [6]. Mercury was analyzed using a Coleman MAS-50 mercury analyzer after the method of Hatch and Ott [7].

Anions, Kjeldahl nitrogen and ammonia were determined using the following procedures from EPA manual EPA 600/14-79-020 [6]; SO_4^{2-}(375.4), F^- (340.2), NO_3^- (353.3), NO_2^- (354.1), PO_4^{3-}(365.3), Kjeldahl-N (351.3) and ammonia-N (350.2). Chloride was analyzed spectrophotometrically according to ASTM method C, p. 364 [8]. All spectrophotometric analyses were performed using a Bausch and Lomb Spectronic 21 spectrophotometer.

Controls were run with all analyses using EPA Quality Control Samples for Trace Metals, Nutrients and Mineral/Physical Analyses. Deionized water was used for standards, and blanks were always determined.

RESULTS AND DISCUSSION

General Chemistry and pH as a Function of Altitude

Precipitation events were collected on an individual basis during the sampling period of April 1979 to August 1980, so that dry deposition was excluded. A total of 52 samples were collected at the valley site and 24 at the mountain site. The pH of the samples was taken immediately after collection and the samples were filtered to remove dust and dirt particles because it was found that neutralization took place as the samples were stored. This effect has been noted elsewhere [9]. The pH values for the mountain and valley stations are shown in Table 1. As can be seen, the average pH (5.24 valley, 5.26 mountain) for all samples is similar at the two sites. This is lower than carbon dioxide equilibrium, which is 5.67. The pH values tended to be lower in the summer than in the winter at the valley site, as has been observed in the northeastern United States, although standard deviations are too large to make the differences statistically significant [1, 10]. A comparison of values obtained during similar time periods in 1979 and 1980 shows a drop in pH values of about 0.3 pH units from 1979 to 1980, indicating that the pre-

Table 1. Average pH Values and Standard Deviations
at the Valley and Mountain Sites[a]

	pH	
	Valley (1400 m) Avg. Precipitation = 20 cm	Mountain (3240 m) Avg. Precipitation = 48 cm
All samples	5.24 ± 0.78 (52)	5.26 ± 0.72 (24)
(May–Sept)	5.12 ± 0.78 (31)	5.23 ± 0.76 (17)
(Oct–April)	5.47 ± 0.76 (21)	5.34 ± 0.67 (7)[b]
(April–Sept 1979)	5.24 ± 1.09 (11)	5.34 ± 0.72 (8)
(April–Sept 1980)	5.02 ± 0.45 (26)	5.15 ± 0.75 (11)
Highest Value	7.56	6.60
Lowest Value	3.66	4.34

[a]The number of samples is shown in parentheses.
[b]Snow samples.

cipitation has become more acidic by a factor of 2. However, the standard deviations are so large that the pH drop is not significant and longer sampling times are required to determine if it is indeed a trend. The pH extremes shown in Table 1 are quite large and are probably a function of both the amount of neutralization occurring and the amount of atmosphere to which the rain is exposed. The extremes at the mountain site were much narrower than those in the valley site, with a factor of 100 difference at the mountain site and a factor of 10,000 difference at the valley site. The high altitude at the mountain site reduces the amount of scavenging time for the rain. Also, the grass and tree cover at the mountain site reduces the blowing dust so that less neutralization occurs. Both of these factors would tend to narrow the gap between the extremes.

Major ion concentrations are shown in Table 2. The predominant cations are calcium and sodium and the major anions are sulfate and chloride.

Very little of the oxidized nitrogen is in the form of nitrite with almost a 10:1 ratio of nitrate-N to nitrite-N. In both sets of analyses, the Σ milliequivalents of cations is close to the anion milliequivalents, indicating an ion balance. Again, the standard deviations are large and statistically there is no difference between the sample sites. However, the smaller values for average major ions indicates a trend to lesser amounts of dissolved solids in precipitation at the higher altitude. This is consistent with less human activity, blowing dust, etc. at the mountain site.

Trace Metals

The average concentration of trace metals in the precipitation at the two stations (38 samples valley site and 10 samples mountain site) is shown in

Table 2. pH Average Concentration and Standard Deviation of Major Ions at Mountain and Valley Sites

	Valley (43 samples)		Mountain (16 samples)	
	Concentration (ppm)	Milliequivalents	Concentration (ppm)	Milliequivalents
Ca	2.0 ± 1.9	0.100	1.7 ± 2.9	0.085
Mg	0.30 ± 0.40	0.025	0.11 ± 0.19	0.009
Na	1.9 ± 2.5	0.083	0.72 ± 1.5	0.031
K	0.54 ± 0.69	0.014	0.27 ± 0.29	0.007
pH	5.24	0.006	5.26	0.005
SO_4	4.6 ± 5.1	0.097	3.4 ± 3.0	0.071
Cl	1.1 ± 0.85	0.031	1.1 ± 1.1	0.031
NO_2/NO_3 as N	0.03 ± 0.04/		0.04 ± 0.04/	
NH_4 as N	0.26 ± 0.27	0.021	0.43 ± 0.31	0.034
HCO_3	0.58 ± 0.49	0.041	1.3 ± 1.9	0.072
F	4.1	0.067	1.8 ± 1.5	0.030
PO_4	0.09 ± 0.08	0.005	0.12 ± 0.10	0.006
	0.11 ± 0.15	0.003	0.54 ± 0.34	0.017
Σ cations		0.269		0.209
Σ anions		0.224		0.189

Table 3. Elements that were not detectable in > 90% of the samples were arsenic, selenium, vanadium, zinc, beryllium and molybdenum. In most instances the average values are similar for both stations, but the elements Ba and Fe appear to be present in greater concentrations at the valley station. This is probably related to the greater amount of blowing dust in the valley. The absence of significant amounts of lead in valley precipitation is indicative of the negligible contribution from automobiles in the region.

Effect of Dry Deposition

After a snowfall in February 1980, it was possible to analyze the effects of dry deposition, because snow remained on the collection funnel for 5 days and only slowly melted. Melt samples were collected during this period and analytical results are presented in Table 4. At the end of 5 days of exposure, the major cations Ca, K and Mg were much higher than at the time of snowfall and the pH had risen, indicating neutralization. The fact that the sulfate values remained unchanged indicates that the dry deposition contained little sulfate. The large calcium ion increase may be due to calcite ($CaCO_3$) in the dry deposition, which neutralized some of the acid to cause the pH rise; the only ion contributed if that were the case would be calcium. This kind of experiment can only be performed under a specific set of weather conditions, which was repeated one other time during the sampling period with the same general results.

Table 3. Average Concentrations and Standard Deviation of Trace Metals in Precipitation

Metals[a]	Valley (38 samples) (ppb)	Mountain (10 samples) (ppb)
Ag	0.18 ± 0.09	0.31 ± 0.37
Ba	19 ± 20	6.5 ± 7.2
Cd	1.4 ± 1.7	1.5 ± 0.9
Cr	1.5 ± 2.7	1.6 ± 2.1
Cu	11 ± 12	11 ± 9
Fe	1.0 ± 4.5	0.3 ± 0.07
Hg	1.1 ± 1.1	0.09 ± 0.7
Pb	6.2 ± 6.9	6.4 ± 6.3
Mn	15 ± 17	14 ± 18
Ni	7.0 ± 8.8	5.4 ± 2.9

[a]Not detected in 90% of samples: As, Se, V, Zn, Be, Mo. Detection limits: As, Se, Mo = 2 ppb; Be = 0.1 ppb; V, Zn = 1 ppb.

Table 4. Concentrations of Major Ions, pH, and SO_4/Ca in Fresh Snow (2/8/80) and Snow Contaminated by Dry Deposition

Date	SO_4	Ca	Na	K	Mg	pH	SO_4/Ca
2/8/80	1.8	0.65	0.95	0.19	0.08	5.93	1.2
2/11/80	1.5	0.34	0.22	0.26	0.06	5.60	1.0
2/12/80	–	0.67	2.5	0.41	0.07	6.15	–
2/13/80	1.4	4.5	0.70	1.3	0.23	6.20	0.13

Table 5. Sequential Rain Sampling During a Single Event Valley Station—August 4, 1980

Time	SO_4	Ca	Na	K	Mg	pH	NO_3 (total as N)	Cl	SO_4/Ca
1510	5.8	2.9	3.6	0.36	1.7	5.03	0.52	1.5	0.8
1515	3.0	2.3	2.5	0.38	0.90	4.99	0.54	1.2	0.5
1520	1.2	0.40	0.11	0.13	0.20	4.24	0.28	0.4	1.3
1525	2.1	0.78	0.25	0.27	0.22	4.34	0.09	0.6	1.1
1530	1.3	0.30	0.17	0.08	0.20	4.18	0.24	–	1.8
1605	1.6	0.36	0.19	0.07	0.20	5.06	0.51	0.4	1.9

Table 6. Sequential Rain Collection During a Single Event Mountain Station

Time	SO_4	Ca	Na	Cl	Mg	K	pH	SO_4/Ca
(8-13-79)								
1406	1.5	0.11	0.05	0.6	<0.02	<0.05	4.35	5.7
1500	1.6	0.13	0.12	1.0	<0.02	<0.05	5.00	5.1
(8-14-79)								
1400	1.5	2.1	0.10	0.6	0.02	0.08	4.42	0.3
1600	2.8	0.49	–	1.1	0.05	0.33	4.74	2.4

Sequential Sample Collection During a Single Event

Sequential sampling during precipitation events usually shows that wash-out predominates during initial phases of the event, resulting in a total ion concentration drop and a pH rise as sampling continues [11, 12]. Analytical results obtained for sequential sampling at the valley site are shown in Table 5. Concentrations of the major cations, SO_4^{2-} and Cl^-, all decreased during the event, indicating that washout was occurring. The pH was higher in the initial sample, dropped slowly, and then rose toward the end of the event. This behavior is different from that ordinarily observed, because the pH is usually low at the beginning of an event when scavenging is greatest [1, 13, 14]. This suggests that the neutralizing effect of airborne particulates is significant in this area. The sulfate-to-calcium ratio rose during the event, while both sulfate and calcium concentrations fell. The calcium concentration dropped more rapidly than the sulfate, which may be due to the calcium being associated with larger particulates arising from blowing dust, while the sulfate may be associated with smaller aerosols, which are less susceptible to washout. This may be the case if the sulfate is formed from SO_2 in the atmosphere. This set of analyses is typical and has been repeated for other events.

Analyses of sequential samples collected at the mountain site are shown in Table 6. The results are more well behaved than sequential samples collected at the valley site, in that pH values are lower at the initial stages than at later stages during an event. There is also no widespread significant drop in major ion concentrations during the event as occurs during valley events. This again indicates the lack of windblown, terrestrial dust at the mountain station as compared to the valley station.

Ion Mole Ratios in Precipitation

Examination of ion mole ratios in precipitation may give clues in locating the sources of the ions as well as their relative importance. For instance, NO_3/SO_4 ratios indicate the relative importance of nitric and sulfuric acids in determining the acidity of precipitation. Ratios for the New Mexico sites and for analyses taken from the literature are summarized in Table 7.

The Cl/Na ratio exhibits the most consistent trends in the literature and calculated values are generally similar to the ratio found in seawater (Cl/Na = 1.1). However, ratio values calculated from Likens et al. [3] for the northeastern United States and average values found in this study are much higher than 1.1 and indicate an excess of chloride. It is difficult to imagine a terrestrial source of chlorine without accompanying cations, although volcanic activity and forest fires are possible contributors. Values for Cl/Na for historic inland precipitation samples collected in 1955–56 by Junge and Werby

Table 7. Ion Mole Ratios for Central New Mexico Compared to Literature Values Calculated

	$\dfrac{SO_4}{Ca}$	$\dfrac{Cl}{Na}$	$\dfrac{NO_3}{SO_4}$
Central New Mexico			
Mountain	3.0	4.5	0.58
Valley	2.8	3.3	0.44
Northeast U.S. [9]	–	0.71	0.88
New Hampshire [3]	6.9	2.7	0.39
Pasadena, CA [15]	6.3	1.2	2.5
Sweden, Inland Unpolluted [1]	1.4	1.4	–
Ireland, Coastal Unpolluted [1]	1.1	1.7	0.02
At Sea, North Carolina [11]	2.5	1.2	0.26
Norway, Coastal Polluted [1]	3.6	1.2	0.66
Norway, Inland Polluted [1]	6.5	1.2	1.0
Seawater Composition [16]	2.7	1.1	–

[17] were calculated from their analysis data to be 0.24 in Albuquerque, New Mexico. The values obtained in this study are an order of magnitude greater and indicate a shift in major ion chemistry. A possible source of excess chlorine may be coal-fired power plants. Chlorine is a common element found in coal, with a 0.1% concentration reported by Klein et al [18]. Analysis of elemental budgets for coal-fired power plants by Klein et al. [18] and by Klein and Andren [19] shows that virtually all of the chlorine present in coal is released as gas. No mention of the fate of this released chlorine is made. It is possible that Cl/Na ratios may be more indicative of coal-fired power plant discharge effects than total sulfur or SO_4/Ca ratios, because there are fewer sources of excess chlorine than of sulfur. This particular aspect deserves more study, especially adjacent to power plants.

The calculated SO_4/Ca ratios show patterns in that the ratios are generally higher in the northeastern United States and in polluted areas in Europe. Again, Junge and Werby [17] obtained much lower values for this study area of 0.21 for SO_4/Ca. Values for this ratio in the New Mexico sampling areas are intermediate among recent values, but strong evidence has been presented in the preceding subsections that a calcium-rich material (probably calcite) may be relatively abundant in terrestrial dust, which would lower the ion ratio on neutralization.

The relative abundance of nitric and sulfuric acids is also intermediate in New Mexico when compared to that calculated from the literature (Table 7). There can only be an insignificant contribution from automobiles to the NO_x content of the atmosphere in central New Mexico because of the very

Table 8. Ion Mole Ratios Calculated on a Total and Seasonal Basis

	Valley Site[a]	Mountain Site[a]
SO_4^{2-}/Ca^{2+}		
Total	2.8 ± 4.2 (39)	3.0 ± 5.9 (17)
May–Sept	3.7 ± 5.3 (20)	1.8 ± 1.4 (12)
Oct–April	1.7 ± 2.3 (19)	5.8 ± 11 (5)
Cl^-/Na^+		
Total	3.3 ± 4.6 (20)	4.5 ± 7.6 (15)
May–Sept	1.3 ± 0.8 (10)	2.6 ± 2.3 (11)
Oct–April	5.4 ± 5.9 (10)	9.7 ± 14.2 (4)
NO_3^-/SO_4^{2-}		
Total	0.44 ± 0.93 (24)	0.58 ± 0.40 (14)
June–Sept	0.63 ± 1.15 (15)	0.70 ± 0.36 (10)
Oct–May	0.13 ± 0.13 (9)	0.27 ± 0.35 (4)

[a]Number of analyses are in parentheses.

small population density in the area. It is probable that the nitrate content of the rain is mainly due to lightning. Evidence for this can be seen by examining Table 8, in which the mole ratios of ions have been calcuated on a seasonal basis. The NO_3/SO_4 ratio is much higher during the summer thunderstorm season (June–September) at both stations than in the winter, when lightning is very rare. Total nitrate occurring during a violent thunderstorm is variable, as can be seen in Table 5. This event produced 4.3 cm of rain in less than an hour, compared to a yearly average of 20 cm for Socorro, so it was a particularly significant event. The predominant acid in the winter appears to be sulfuric and in the summer the nitric acid contribution is high. There is no difference between averages based on a t-test at 90% confidence limits, but the trends seem evident.

The SO_4/Ca ratio shows opposing patterns from summer to winter at the two sites, with more sulfate present at the valley site in the summer and more sulfate at the mountain site in the winter. There were five samples from the mountain site in the winter and they were all snow samples. Three snow samples collected at the valley site during this same period gave an average SO_4/Ca ratio of 0.9, which is close to the 1.8 ratio for all samples during this same time period. The apparent high ratios at the mountain site again may be due to less dust neutralizing the acid.

The snow samples from the mountain site also exhibited high Cl/Na ratios (Table 8) and the trends for the mountain and valley stations were the same on a seasonal basis. The Cl/Na ratio was higher in the winter than in the summer. These ion ratios require more study and may be related to terrestrial dust, aerosols or gases from coal combustion. Examination of dust and aerosols in the region is presently underway.

SUMMARY

The pH values of precipitation in the high desert and mountain region in south-central New Mexico averaged 5.25 over an 18-month period, indicating more acidity than contributed by CO_2 alone. Also, the acidity has increased by a factor of 2 over similar periods in 1979 and 1980, although there is too much scatter in the data to state that the change is significant. Longer-term sampling is necessary to judge trends indicated in this study.

Terrestrial dust has a large neutralizing effect on precipitation events in the region, especially in the initial stages. This effect is more evident in thunderstorms when blowing dust is present. Values obtained for Cl/Na and SO_4/Ca ratios are much greater than those reported for 1955–1956 in the same region, indicating an increase in the Cl and SO_4 anions relative to major cations.

Sulfuric acid appears to predominate in precipitation in the area, except in the summer, when lightning appears to produce a larger nitric acid component.

ACKNOWLEDGMENTS

C. M. Jensen would like to express appreciation to the Research and Development Division of New Mexico Tech for partial support during this project. The authors wish to express thanks to the Director of Langmuir Laboratory, Mr. Charles Moore, for his cooperation in making the facilities available and to the laboratory personnel for their aid in sample collection.

REFERENCES

1. Likens, G. E., et al. "Acid Rain," *Scientific Am.* 241:43–51 (1979).
2. Likens, G. E. "Acid Precipitation," *Chem. Eng. News* (November 27, 1976), pp. 29–37.
3. Galloway, J. N., G. E. Likens and E. S. Edgerton. "Acid Precipitation in the Northeastern United States," *Science* 194:722–724 (1976).
4. Lewis, Jr., W. M. and M. C. Grant, "Acid Precipitation in the Western United States," *Science* 207:176–7 (1980).
5. Kleber, E., A. Dasti and A. Gakner. "The Effects on the Environment of Fossil Fuel Usage by Electric Utilities," presented at the 180th National American Chemical Society Meeting, Environmental Division, Las Vegas, August 1980.

6. "Methods for the Analysis of Water and Wastes, U.S. EPA 600/14-79-020 (1979).

7. Hatch, W. R. and W. L. Ott, "Determination of Sub-Microgram Quantities of Mercury by Atomic Absorption Spectroscopy," *Anal. Chem.* 40:2085 (1968).

8. *Annual Book of ASTM Standards*, American Society for Testing and Materials, Part 31, Water, Philadelphia, PA (1980).

9. Pack, D. H. "Precipitation Chemistry Patterns: A Two-Network Data Set," *Science* 208:1143-1145 (1980).

10. Wolff, G. T., et al. "Acid Precipitation in the New York Metropolitan Area: Its Relationship to Meteorological Factors," *Environ. Sci. Technol.* 13:209-212 (1979).

11. Gambell, A. W. and D. W. Fisher. "Occurrence of Sulfate and Nitrate in Rainfall," *J. Geophys. Res.* 69:4203-4210 (1964).

12. Brezonik, P. L., E. S. Edgerton and C. D. Hendry. "Acid Precipitation and Sulfate Deposition in Florida," *Science* 208:1027-1029 (1980).

13. Forland, E. J., "A Study of the Acidity in the Precipitation in Southwestern Norway," *Tellus* 25:291-298 (1973).

14. Forland, E. J. and Y. T. Gjessing. "Snow Contamination from Washout/Rainout and Dry Deposition," *Atmos. Environ.* 9:339-352 (1975).

15. Liljestrand, H. M. and J. J. Morgan, "Chemical Composition of Acid Precipitation in Pasadena, Calif.," *Environ. Sci. Technol.* 12:1271-1273 (1978).

16. Krauskopf, K. B. *Introduction to Geochemistry*, 2nd ed., (New York: McGraw-Hill Book Co., 1979), p. 263.

17. Junge, C. B. and R. T. Werby, "The Concentration of Chloride, Sodium, Potassium, Calcium and Sulfate in Rain Water Over the United States," *J. Meteor.* 15:417-425 (1950).

18. Klein, D. H., et al. "Pathways of Thirty-Seven Trace Elements Through Coal-Fired Power Plant," *Environ. Sci. Technol.* 9:975-979 (1975).

19. Klein, P. H., A. W. Andren and N. E. Bolton. "Trace Element Discharges from Coal Combustion for Power Production," *Water Air Soil Poll.* 5:71-77 (1977).

CHAPTER 7

CHEMICAL SOURCE, EQUILIBRIUM AND KINETIC MODELS OF ACID PRECIPITATION IN SOUTHERN CALIFORNIA

Howard M. Liljestrand and James J. Morgan

Environmental Engineering Science
California Institute of Technology
Pasadena, California 91125

INTRODUCTION

The acid precipitation phenomenon in northern Europe and the northeastern United States appears to be a consequence of long-range transport of anthropogenic emissions of air pollutants [1, 2]. The predominant acid is sulfuric acid and the pH is controlled by the interaction of acids (sulfuric and nitric) with bases (ammonia and soil dust) [3]. Emissions of acidic pollutants can result in acid precipitation on a local scale, even though the regional precipitation may be neutral or alkaline [4-6].

The Los Angeles basin offers an interesting environment for study of precipitation acidity since long-range transport of acidity to Los Angeles is not expected during storms and high local emissions of NO_x lead to nitric acid as the predominant acid in the West [6-9]. Precipitation has been sampled at Pasadena, California from February 1976 to August 1979, and eight other sites in the Los Angeles area have been sampled for shorter intervals. The acidity and chemical composition of the precipitation are examined to relate primary sources, air quality and rainwater quality.

103

EXPERIMENTAL METHODS

Rainfall samples for inorganic analysis were obtained in an all-plastic collector which opened automatically at the beginning of precipitation and collected sequential 1/4-in. rainfall samples throughout a storm. Check-valve floats preserved the integrity of samples prior to analysis. An all-glass and metal collector was used to collect samples for organic analyses [10].

pH was determined electrometrically, using an Orion 801A digital pH/mV meter with glass electrode-double junction reference electrode cell. The cell was calibrated with dilute solutions of strong acids to minimize junction potential effects. Samples were titrated with standard base of concentration 0.001 N in a closed-jacketed beaker in order to determine acidity. Initial pH was recorded, then high-purity nitrogen or argon was bubbled through the sample for 30 minutes to remove dissolved CO_2. The titration was carried out under a nitrogen/argon atmosphere. Gran functions from the titration data were plotted to distinguish strong and weak acid components [7,11,12].

Results of such titrations show that acidities in the Los Angeles area samples consist of free acidity (H^+), dissolved carbon dioxide acidity ($H_2CO_3{}^*$), ammonium ion acidity ($NH_4{}^+$) and small amounts of weak acids such as organic acids (RCOOH) and hydrolyzable metals (e.g., Al^{3+}, Fe^{3+}). The net acidity, defined with respect to a pH datum of 5.65 (water in equilibrium with normal atmospheric CO_2), is essentially identical to free acidity ($[H^+]$) for southern California samples. Thus, laboratory pH measurements and acidity titrations yield concordant results.

A Dionex Model 10 ion chromatograph was used for determining anion concentrations of chloride, nitrate, sulfate, fluoride, bromide, nitrite and orthophosphate. A 100-μL sample loop was used with 0.003 M $NaHCO_3$/ 0.0024 M Na_2CO_3 eluant. For very dilute samples an anion concentrator column was used. In addition, fluoride was determined by the SPADNS method given in *Standard Methods* [13], chloride was determined by the method of Florence and Farrar [14] and total phosphate was determined spectrophotometrically [13].

Metals were determined by atomic absorption spectrophotometry (AAS). The following metals were determined by flame or carbon rod AAS, depending on sample concentration: sodium, potassium, calcium, magnesium, iron, aluminum, manganese, nickel, lead and zinc. $NH_4{}^+$ (ammonium) was determined in two ways: by means of an NH_3 electrode together with pH measurement, and by the phenate standard method [13].

Silica (orthosilicic acid) was determined by the spectrophotometric method described by Strickland and Parsons [15]. Conductivity was determined on each sample using a conductivity meter. Total organic carbon was determined with a Dohrmann Envirotech instrument. Suspended solids were determined gravimetrically after collection on a Whatman GF/C filter.

The chemical data obtained in this study were subjected to three major consistency checks: (1) charge balance for all significant cations and anions determined; (2) conductivity balance between measured conductivity and that calculated from concentrations of all significant cations and anions; (3) comparison of strong acidity and weak acidity obtained from acidity titration curves with the corresponding measured values of hydrogen ion [H^+], ammonium ion [NH_4^+], and other weak acids. Each of these tests showed excellent internal consistency in the analytical results obtained at each location.

CHEMICAL SOURCE RELATIONSHIPS

The chemical composition of the precipitation, given in Table 1, is assumed to be a linear combination of the compositions of the major sources: sea salt, soil dust, ammonia, mobile sources and stationary sources of fossil fuel combustion. The mass ratio of the ith chemical species to the total residue in rainwater (X_i) is given by

$$X_i = \sum_j \alpha_{ij} X_{ij} A_j \qquad (1)$$

where X_{ij} is the mass ratio of the ith chemical species for emissions from the jth source, and α_{ij} is the fractionation factor for the ith element from the jth source. A_j is the source strength or mass fraction of total residue from the jth source to the rainwater. Similar models have been used for aerosols [16] and precipitation [3].

Since there are more chemical species than sources, the least-squares best fit of the overdetermined system of equations is found to give the best estimate of each source contribution, shown in Figure 1. There are still uncertainties in this type of mixing model calculation because of assumptions concerning the major sources and fractionation of species between the source and sink [17]. Slinn et al. [18] have reviewed the average fractionation factors for precipitation scavenging of chemical species in the aerosol and gas phases.

The kinetics of precipitation scavenging are reflected in the fractionation factors (α_{ij}). Since more than twice as much NO_x as SO_2 is emitted on an equivalent basis in southern California [19], nitric acid is the predominant acid, in contrast to the Northeast, where emissons of SO_2 are greater than emissions of NO_x and, consequently, sulfuric acid is found to be the predominant acid in precipitation. The precipitation-weighted equivalent ratio of nitrate to non-sea-salt sulfate is 1.1 in southern California. The ratio is quite variable, since sulfuric acid is the predominant acid near the coastal

Table 1. Precipitation-Weighted-Mean Concentrations for the 1978–1979 Hydrologic Year [10]

	H^{+a} μN	NH_4^+ μN	Na^+ μN	K^+ μN	Ca^{2+} μN	Mg^{2+} μN	Cl^- μN	NO_3^- μN	SO_4^{2-} μN	Total NO_2^- μN	Br^- μN
Long Beach	29	14	37	0.96	8.8	11	42	19	51	0.5	0.21
Westwood	30	21	29	1.5	10	8.8	33	27	42	0.8	0.42
Central Los Angeles	32	36	34	4.9	15	11	40	34	56	0.4	0.57
Pasadena	39	21	24	1.7	6.7	7.2	28	31	39	0.5	0.25
Mt. Wilson	10	36	26	1.7	9.3	6.6	28	23	40	<0.4	0.17
Azusa	22	36	24	1.7	15	8.0	28	44	38	0.5	0.37
Wrightwood	13	0.86	4.1	0.24	3.9	1.7	5.0	11	7.2	<0.4	0.032
Riverside	11	34	25	2.4	17	8.4	30	33	33	0.9	0.17
Big Bear	3.8	7.5	4.2	0.44	9.3	7.5	5.2	17	6.5	<0.4	0.019

[a]Precipitation-weighted-mean concentrations are defined by $\Sigma P_i C_i / \Sigma P_i$ where P_i and C_i are the precipitation and concentration of a conservative species for the ith sample, respectively. Acidity is the conservative species averaged to give the precipitation-weighted-mean hydrogen ion concentration [H^+]. For samples with pH controlled by strong acids, the acidity is approximately equal to the hydrogen ion concentration.

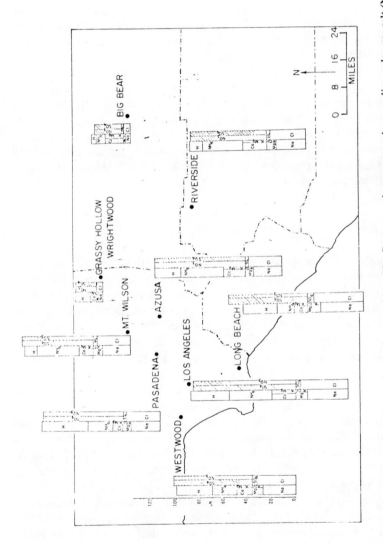

Figure 1. Source contributions to the mean chemical composition. The five assumed sources are, in ascending order, sea salt (NaCl + MgSO₄), soil dust (predominantly Ca) neutralized acids, ammonia neutralized acids, nitric and sulfuric acids from mobile sources (open), and sulfuric and nitric acids from stationary sources (crosshatched area) [10].

stationary sources, and nitric acid is the predominant acid at inland sites. Since nitrogen oxides are scavenged less effectively than sulfur dioxide [18], the equivalent ratio of nitric to sulfuric acid in precipitation is less than the equivalent ratio of NO_x to sulfur oxides from pollutant sources.

Several trends are evident from the acid-base source characterization. First, the net acidity can be viewed as the neutralization of strong acids (nitric and sulfuric) by bases (ammonia and soil dust) at all sampling sites. Second, non-sea-salt sulfate decreases inland and is lowest in the mountains. This trend coincides with the distribution of sulfur oxide sources [20]. Third, the nitrate to non-sea-salt sulfate ratio increases from coastal to inland sites.

The ammonium and soil dust trends are less obvious. The higher soil dust contributions at central Los Angeles and Azusa reflect construction and earth moving acitvity near sampling sites. The high ammonium concentration at Azusa is expected as a result of dairy feedlots in the area. Agricultural sources are more pronounced at Riverside, as shown by higher ammonium and soil dust contributions.

EQUILIBRIUM RELATIONSHIPS

Equilibria relating air and water quality are given in Table 2, along with pertinent thermodynamic data [21-25]. Hales [26] has reviewed the kinetic limitations of thermodynamic models. For reactive gaseous species, i.e., those which transfer and react rapidly in water, the kinetic limitations may be of less consequence than the uncertainty about the gaseous concentrations, which are near the limits of detection during precipitation.

While the uncertainties are high, total sulfite ($[SO_2 \cdot H_2O] + [HSO_3^-] + [SO_3^{2-}]$) could account for a significant fraction of the observed excess or non-sea-salt sulfate. No chemical preservatives were added to samples to prevent oxidation of sulfite to sulfate prior to analysis. Total sulfite concentrations could account for 10 to 35% of the excess sulfate at central Los Angeles and Riverside, respectively. These results are comparable to those of Hales and Dana [27] for the Northeast. Aqueous-phase oxidation of sulfite to sulfate during precipitation would increase the contribution of SO_2 gas scavenging processes to the formation of acid precipitation.

The equilibrium dissolution of NO and NO_2 gases, to form nitrous acid and nitrite (reactions 2.4 and 2.5) accounts for observed total nitrite ($[HNO_2] + [NO_2^-]$) concentrations. While NO_x dissolution to form nitrite is kinetically fast [28], the overall reaction has a low thermodynamic driving force in acidic solutions. Thus, equilibrium is approached, even though total nitrite concentrations are a minor component in precipitation acidity.

Table 2. Thermodynamic Data for pH-Controlled Dissolution and Dissociation Reactions [21-25]

	Reaction	$\Delta G^\circ_{298.15}$ (kcal/mole)	$\Delta H^\circ_{298.15}$ (kcal/mole)	$K_{eq\,(298.15)}$ [a]
2.1	$SO_{2(g)} + H_2O_{(L)} = SO_2 \cdot H_2O_{(aq)}$	-0.130	-6.247	1.245
2.2	$SO_2 \cdot H_2O_{(aq)} = H+ + HSO_3^-$	2.578	-4.161	1.29×10^{-2}
2.3	$HSO_3^- = H^+ + SO_3^{2-}$	9.850	-2.23	6.014×10^{-8}
2.4	$NO_{(g)} + NO_{2(g)} + H_2O_{(L)} = 2\ HNO_{2(aq)}$	-2.857	-18.185	1.243×10^2
2.5	$HNO_{2(aq)} = H^+ + NO_2^-$	4.488	3.5	5.129×10^{-4}
2.6	$HNO_{2(g)} = H^+ + NO_2^-$	-2.48	-9.35	6.577×10^1
2.7	$2\ NO_{2(g)} + H_2O_{(L)} = HNO_{2(aq)} + H^+ + NO_3^-$	-7.717	-25.565	4.541×10^5
2.8	$HNO_{3(g)} = H^+ + NO_3^-$	-8.92	-17.46	3.460×10^6
2.9	$NH_{3(g)} + H_2O_{(L)} = NH_4OH_{(aq)}$	-2.41	-8.17	5.844×10^1
2.10	$NH_4OH_{(aq)} = NH_4^+ + OH^-$	6.503	8.65	1.709×10^{-5}
2.11	$H_2O_{(L)} = H^+ + OH^-$	19.093	13.345	1.008×10^{-14}

[a]Equilibrium constants for temperatures other than 298.15° K can be estimated by the Van't Hoff relation, $\log K = \log K_{298.15} - \Delta H^\circ_{298.15} / 2.3\ R\ (1/T - 1/298.15)$.

In addition, Table 3 presents gaseous pollutant partial pressures which are in equilibrium with rainwater. These air concentrations are applicable only for precipitation conditions and only for thermodynamic equilibrium of pH-controlled dissolution reactions. Nitrate formation from NO_2 (reaction 2.7) is kinetically slow, and equilibrium is not attained [29].

The partial pressures of ammonia (a strong base) and nitric acid (a strong acid) are below the limits of detection of monitoring techniques. At high relative humidity and water content in the air during precipitation, these very soluble gases will condense/dissolve into hygroscopic aerosol droplets as well as raindrops. Low partial pressures do not necessarily imply removal from the atmosphere, but may reflect partitioning of gases into the aerosol phase as well as steady-state concentrations between release/formation in the atmosphere and scavenging.

Precipitation is poorly buffered with respect to very soluble strong acid and base gases. The trend of increasing partial pressure of ammonia from coastal to inland sites is evident. The highest partial pressure of nitric acid occurs at Pasadena, reflecting source distributions as well as reaction kinetics. Similar values of the partial pressure of ammonia during precipitation have been calculated by Dawson [30] and Lau and Charleson [31] using the same equilibrium model. A steady-state nonequilibrium model proposed by Hales and Drewes [32] indicates the partial pressure of ammonia to be a factor of 40 higher than estimated by equilibria 2.9-2.11.

Equilibrium with respect to oxidation-reduction reactions can be determined if oxidant and reductant concentrations are known. Direct measurements of oxidant or reductant concentrations worldwide are limited. Table 4 summarizes the identities and concentrations of oxidants and reductants measured or calculated to be in rainwater. Calculated values are based on gas-phase measurements with the assumption of equilibrium solubility of the gaseous species. This assumption is not adequate for reactive species and represents an upper bound for the actual concentration.

The magnitude of an oxidant/reductant concentration may belie the importance of its reactions. In the steady-state of dissolution and reaction, low concentrations can have a high thermodynamic driving force, presented in Table 5 in the form of half-cell potentials. The standard half-cell potentials ($E°$) are given as well as typical potentials for precipitation concentrations (from Table 4). By applying the range of concentrations found in Table 4 with assumed pH values (2.5-7.5, median 4.3) to the Nernst equation,

$$E = E° + \frac{RT}{nF} \ln \frac{\{Ox\}}{\{Red\}} \qquad (2)$$

potentials for each half-cell reaction are estimated.

Table 3. Average Values Calculated from Equilibrium Relationships

	Long Beach	Westwood	Los Angeles	Pasadena	Azusa	Riverside
Total Sulfite[a] (μN)	5.5	5.3	4.9	4.1	7.2	7.4
Non-Sea Salt SO_4^{2-} (μN)	46.5	39.0	51.4	35.6	34.9	30.4
$\dfrac{\text{Total Sulfite}}{\text{Non-Sea Salt } SO_4^{2-}}$	0.12	0.14	0.10	0.12	0.21	0.24
Total Nitrite[b] (μN)	0.56	1.0	0.41	0.67	0.50	1.3
P_{HNO_3}[c] ATM x 10^{15}	0.051	0.075	0.10	0.11	0.09	0.034
P_{NH_3}[d] ATM x 10^{12}	1.0	1.6	2.5	1.2	3.7	7.0

[a] Total Sulfite $= [SO_2 \cdot H_2O] + [HSO_3^-] + [SO_3^{2-}]$
$$= P_{SO_2}K_H + P_{SO_2}K_HK_1/[H^+] + P_{SO_2}K_HK_1K_2/[H^+]^2 \text{ from 2.1 to 2.3.}$$

[b] Total Nitrite $= [HNO_2] + [NO_2^-]$
$$= \sqrt{K_HP_{NO}P_{NO_2}} + \sqrt{K_HP_{NO}P_{NO_2}}\ (K_1/[H^+]) \text{ from 2.4 to 2.5.}$$

[c] $P_{HNO_3} = [H^+][NO_3^-]/K_H$ from 2.8.

[d] $P_{NH_3} = [NH_4^+] K_w/[H^+] K_HK_b$ from 2.9 to 2.11.

Table 4. Calculated and Measured Oxidant/Reductant Concentrations

Species	Range	Mean/Median	Assumptions
O_3	$(0.85–8.5) \times 10^{-10} M$	$3.4 \times 10^{-10} M$	Equilibrium solubility with P_{O_3} = 5–50 ppbv
H_2O_2	$(2.94–4670) \times 10^{-8} M$	$2.35 \times 10^{-8} M$	Measured at Claremont, CA [33]
O_2	$(2.66–3.91) \times 10^{-4} M$	$3.12 \times 10^{-4} M$	Equilibrium solubility with P_{O_2} = 0.21 ATM
SO_2	$(0.17–13.0) \times 10^{-6} M$	$5.0 \times 10^{-6} M$	Equilibrium solubility with P_{SO_2} = 5–20 ppbv
HNO_2	$(1.1–14.0) \times 10^{-7} M$	$3.9 \times 10^{-7} M$	Measured at Pasadena, CA
NO_3^-	$(1.1–190) \times 10^{-5} M$	$7.5 \times 10^{-5} M$	Measured at Pasadena, CA [7]
NO	$(1.93–38.6) \times 10^{-11} M$	$9.65 \times 10^{-11} M$	Equilibrium solubility with P_{NO} = 10–200 ppbv
NO_2	$(1.9–38.0) \times 10^{-10} M$	$9.5 \times 10^{-10} M$	Equilibrium solubility with P_{NO_2} = 10–200 ppbv

Table 5. Half-Cell Reduction Potentials

	Half-Cell Reaction	$E°$ (V)	E in rainwater (V)	
			Range	Median
5.1	$O_3 + 2H^+ + 2e^- = O_2 + H_2O$	2.04	1.38–1.73	1.62
5.2	$H_2O_2 + 2H^+ + 2e^- = 2H_2O$	1.78	1.11–1.50	1.42
5.3	$O_2 + 4H^+ + 4e^- = 2H_2O$	1.27	0.78–1.07	0.97
5.4	$NO_2 + H^+ + 1e^- = HNO_2$	1.20	0.78–0.97	0.83
5.5	$NO_2 + 2H^+ + 2e^- = NO + H_2O$	1.01	0.57–0.92	0.79
5.6	$HNO_2 + H^+ + e^- = NO + H_2O$	1.14	0.34–0.89	0.73
5.7	$NO_3^- + 4H^+ + 3e^- = NO + 2H_2O$	1.00	0.39–0.86	0.68
5.8	$NO_3^- + 3H^+ + 2e^- = HNO_2 + H_2O$	0.94	0.37–0.82	0.58
5.9	$O_2 + 2H^+ + 2e^- = H_2O_2$	0.77	0.38–0.74	0.57
5.10	$NO_3^- + 2H^+ + 1e^- = NO_2 + H_2O$	0.67	-0.01–0.79	0.46
5.11	$SO_4^{2-} + 4H^+ + 2e^- = H_2SO_3 + H_2O$	0.17	0.37–0.96	0.58
5.12	$2H^+ + 2e^- = H_{2(g)}$	0.00		

Table 5 emphasizes that a number of oxidants are stronger than oxygen in rainwater. Since the mechanism of oxygen reduction proceeds through hydrogen peroxide (reaction 5.9, a two-electron process rather than reaction 5.3, a four-electron process), N(III), N(IV), N(V), hydrogen peroxide and ozone are all stronger oxidants than the O_2/H_2O_2 half-cell. The most abundant oxidant, oxygen, does not provide the highest oxidation potential, and all the above oxidants are strong enough to oxidize S(IV) to S(VI).

In addition, Table 5 implies several half-cell reactions may be coupled and approach oxidation-reduction equilibrium, while others do not approach equilibrium. Similar average half-cell potentials and overlap in the range of potentials indicate equilibrium is approached for several of the NO_x reactions. This result may be coincidental, in that many oxidation-reduction reactions are not coupled due to kinetic limitations. The kinetic limitations of reaction 2.7 are manifest in the disparity between the oxidation-reduction potentials of half-cells 5.4 and 5.10.

STATISTICAL RELATIONSHIPS

A number of linear and nonlinear correlations as well as multiple regression techniques have been used to model rainwater concentrations. To a first approximation, all rainwater concentrations tend to be correlated with one another, and components tend to be correlated with rainfall rate [6, 30, 34]. This "dilution" effect is predicted by physical-chemical scavenging models for condensing/evaporating raindrops [35] and for scavenging efficiency changes with raindrop-size-distribution and precipitation intensity [36]. The exponential washout of air pollutants also predicts decreasing rainwater concentrations with increasing total precipitation, as has been noted for samples taken within a storm [37], samples for separate storms [38] and for aggregates of multiple storms of variable rainfall [39].

A second statistically significant relationship often found is that between $[H^+]$ and the other major ions, which is predicted by the charge balance. The difference in anion and metal plus ammonium ion equivalent concentrations should approximately equal $[H^+]$ for the charge balance to be satisfied in acid precipitation. Correlations have been noted even when sodium and chloride concentrations (often approximately equal as in seawater) and ions of relatively low concentration (potassium, magnesium and sometimes ammonium) are excluded [40].

The third type of relationship confirmed by statistical methods is the common source of species, which is predicted by the chemical source balance described above. Lodge et al. [41] used factor analysis to distinguish soil dust, sea salt and ammonia sources. Ion ratios in northern California precipitation agreed with those predicted from seawater concentrations [6]. In the Puget Sound area, Knudsen et al. [4] used an elegant method of factor analysis to identify urban and industrial (smelter) plumes in rainwater concentrations.

The fourth type of statistically significant relationship often found is between emissions and wet deposition. Vermeulen [42] found linear correlations of acidity ($[H^+]$) with sulfur dioxide and nitrogen oxides emissions for several sites in The Netherlands. A linear relationship between emissions and wet fluxes was found for lead by Lazrus et al. [43] by comparing gasoline consumption with wet lead flux.

Since the charge balance and source-precipitation relationships have been addressed above, multiple-stepwise linear regression [44] was performed on air quality, water quality and atmospheric data for Pasadena, to identify variables important to the physical-chemical scavenging mechanisms. Sulfate as well as nitrate plus nitrite were treated as dependent variables (y) to find the least squares best fit with an equation of the form

$$y = b_o + b_1 x_1 + \ldots + b_{n-1} x_{n-1} \qquad (3)$$

The independent variables (x_i) were: ground level ambient air concentrations of nonmethane hydrocarbons, ozone, carbon monoxide, sulfur dioxide, nitric oxide and nitrogen dioxide during the collection of rainwater samples; temperature, relative humidity and particulate matter. Additional independent variables were precipitation intensity, $[H^+]$, $P_{NO_2}/[H^+]$ and $\sqrt{P_{NO}P_{NO_2}}/[H^+]$. The last two variables were chosen as indicators of nitrogen oxide gas-liquid equilibria (reactions 2.7, 2.5 and 2.4). Finally, rainwater sulfate was an independent variable for rainwater nitrate plus nitrite and vice versa.

For samples collected from January 1978 to April 1979, which were predominantly winter rains, rainwater nitrate and nitrite were dependent on four variables at the 95% confidence level, shown in Table 6. The variance ratio is a test to determine if the variance in the dependent variable explained by the independent variable is greater than that caused by a random variable. Thus, rainwater nitrate and nitrite was more than just randomly dependent on precipitation intensity, ozone, nitric oxide and nitrogen dioxide divided by hydrogen ion concentration.

While multiple regression analysis does not determine the mechanism, these independent variables may represent the following physical-chemical steps in scavenging. Precipitation intensity reflects the dilution effect and scavenging efficiency changes with raindrop size. Ozone may reflect: (1) the ratio of NO_2 to NO, (2) nitric and nitrous acids in the gas/aerosol phase(s) or (3) oxidation by an oxidant stronger than oxygen during the scavenging process. Higher P_{NO_2}/P_{NO} would give both higher equilibrium concentrations of nitrate and nitrite and faster kinetics toward equilibrium. A higher oxidation state for the NO_x as reflected by a higher ozone concentration would enhance the scavenging kinetics of a given NO_x concentration. P_{NO} and $P_{NO_2}/[H^+]$ reflect higher thermodynamic driving forces for precipitation scavenging.

Table 6. Air Quality and Meteorological Parameters with Significant Variance Ratios for Multiple Linear Regression for $[NO_2^-] + [NO_3^-]$ in Pasadena Rainwater

Data Set	Independent Variable	Confidence Level
1/78–4/79	Precipitation intensity	>99%
Winter Storms	Ozone	>99%
n = 198	Nitric oxide	>99%
	$P_{NO_2}/[H^+]$	>95%
2/76–9/79	Temperature	>99%
All Samples	Nitrogen dioxide	>99%
n = 268	Nitric oxide	>99%
	Relative humidity	>99%
	$P_{NO_2}/[H^+]$	>99%
	Carbon monoxide	>99%
	$[H^+]$	>99%
	Sulfur dioxide	>99%
	Ozone	>99%
	$[SO_4^{2-}]$ in rainwater	>95%
1/78–4/79	Hi-Vol lead	>99%
Samples Collected	Particulate matter (1 hr)	>95%
During Hi-Vol Sampling	Ozone	>95%
n = 19	Temperature	>95%

For samples collected from February 1976 to September 1979 (all of the Pasadena samples) which include fall, winter, spring and summer storms, additional variables significantly explained variation in nitrite and nitrate concentration. Temperature and relative humidity could reflect evaporation/condensation effects and the sulfate in rainwater also represents the dilution effect. Alternatively, temperature, $[H^+]$ and sulfate (sulfuric acid) in rain could reflect acid catalysis of NO_x scavenging with faster kinetics at higher temperature. Again, P_{NO}, P_{NO_2} and $P_{NO_2}/[H^+]$ provide higher thermodynamic driving forces for NO_x scavenging, and ozone may reflect higher oxidation states for enhanced scavenging. P_{CO} may reflect the automotive source as discussed below.

Twenty-four-hour Hi-Vol lead, sulfate, nitrate and total suspended particulates were also considered as independent variables that may influence rainwater concentrations. Aerosol is sampled every six days in Pasadena by the South Coast Air Quality Management District. As chance would have it, aerosol was collected during only 19 rainy days. Precipitation samples were collected over periods of 1-2 hr during the 24 hours the Hi-Vol aerosol samples were collected.

Aerosol parameters and rainwater quality would not be expected to be well correlated, because of the different averaging periods. Surprisingly, aerosol lead concentration has the highest variance ratio for both rainwater nitrite plus nitrate and sulfate concentrations. Lead and carbon monoxide

Table 7. Air Quality and Meteorological Parameters with Significant Variance Ratios for Multiple Linear Regression for $[SO_4{}^{2-}]$ in Pasadena Rainwater

Data Set	Independent Variable	Confidence Level
1/78–4/79	Nitric oxide	>99%
Winter Storms	Ozone	>99%
n = 198	Precipitation intensity	>95%
2/76–9/79	Nitric oxide	>99%
All Samples	Ozone	>99%
n = 268	Precipitation intensity	>99%
	$[H^+]$	>99%
	Temperature	>99%
	Sulfur dioxide	>99%
	Relative humidity	>95%
1/78–4/79	Hi-Vol lead	>99%
Samples Collected	Particulate matter (1 hr)	>95%
During Hi-Vol Sampling	Ozone	>95%
n = 19	Temperature	>95%

are both indicators of the automotive source. Other variables that had significant variance ratios (>95% confidence level) for rainwater nitrite and nitrate as well as rainwater sulfate in this nineteen-sample subset were ozone, 1-hr averages of particulate matter and temperature.

The similarities between the variables influencing rainwater sulfate and nitrite plus nitrate concentrations are due to similar sources, transport and scavenging processes. Rainwater sulfate for the predominantly winter samples (January 1978 to April 1979) were significantly dependent on three variables: nitric oxide, ozone and precipitation intensity, as shown in Table 7. Rainwater sulfate for all Pasadena samples was also significantly dependent on temperature, $[H^+]$, relative humidity and sulfur dioxide.

The predictive capability of results of the multiple linear regression (Equation 3) is still qualitative. The best fit often calculates rainwater concentrations that are high or low by a factor of two from the actual concentrations. This error could be due to the exclusion of important variables, the limitations of using ground-level measurements to describe air quality, and the exclusion of environmental conditions upwind of the sampling site. This exercise identifies which environmental variables may be important to kinetic model of precipitation acidity.

CHEMICAL KINETICS

Nitric acid production during precipitation proceeds through three competing mechanisms: gas-phase oxidation followed by dissolution into raindrops, dissolution into aerosol followed by aerosol scavenging by raindrops, and aqueous-phase oxidation in raindrops. The last mechanism appears to be negligible for uncatalyzed reactions. Table 8 summarizes the reported kinetic rate laws for aqueous NO_x oxidation processes. In addition, reaction times required to form observed rainwater concentrations are given for each reaction, assuming typical reactant concentrations as given in Table 4. In each case, reaction times are much longer than droplet lifetimes.

Dissolution into aerosol droplets followed by precipitation scavenging of aerosols suffers the same kinetic limitations as the dissolution and reaction in rainwater. However, the mean age of aerosols at the beginning of storms is sufficiently large that equilibrium NO_x dissolution into aerosols is attained prior to precipitation [51]. The overall contribution of this mechanism (aerosol scavenging) to nitrate concentrations in precipitation can be estimated by washout ratios (r),

$$r = c/x \qquad (4)$$

where c/x is the dimensionless ratio of rainwater concentration (c) to air concentration (x) [52, 53]. The median nitrate aerosol concentration in Pasadena during precipitation is 6.9 $\mu g/m^3$. Assuming the washout ratio for nitrate aerosol is approximately 0.19×10^6 [18, 52], the nitrate concentration in rainwater due to aerosol scavenging should be $\sim 21 \mu M$, or about two-thirds of the average nitrate concentration in Pasadena precipitation. Since the uncertainty in the washout ratio (r) is high, the uncertainty in the contribution of aerosol scavenging to nitrate concentrations in rainwater is also high. The turnover time for aerosols during precipitation would have to be about 50 minutes for a steady state between aerosol formation and aerosol scavenging to be maintained. This time is again less than the reaction times required for uncatalyzed aqueous reactions, as given in Table 8.

The third mechanism, gas-phase reactions followed by dissolution, can account for the required rate of aerosol formation as well as the remaining rainwater nitrate. Free-radical chain reactions in the gas phase [54, 55] predict higher gas-phase partial pressures of N(V) (N_2O_5, HNO_3) than those given in Table 3 for equilibrium dissolution into rainwater. Thus gas-phase reactions followed by gas-to-particle conversion (aerosol formation) and scavenging of gases and aerosols appears to be the process of nitric acid formation. The oxidation step appears to occur in the gas phase.

In contrast, aqueous-phase oxidation appears more important than gas-phase oxidation processes for sulfuric acid formation. The aqueous oxidation may occur in both the aerosol and raindrop phases. Penkett et al. [56] have summarized three competing aqueous phase mechanisms: oxidation by oxygen, predominant at basic pH or with catalysts; oxidation by ozone, predominant at slightly acidic pH; and oxidation by hydrogen peroxide, pre-

Table 8. Aqueous-Phase NO_X Oxidation Kinetic Rate Laws

	Reaction	Kinetic Rate Law [45–50]	Reaction Time in Rainwater to Form $10\ \mu M\ NO_3^-/0.5\ \mu M\ NO_2^-$
8.1	$4\,NO + O_2 + H_2O \rightarrow 4\,HNO_2$	$\dfrac{d[NO]}{dt} = \dfrac{8.8 \times 10^6\,[NO]\,2\,[O_2]}{M^2 \cdot s}$	3.9×10^{10} sec
8.2	$4\,NO + 3\,O_2 + 2\,H_2O \rightarrow 4\,HNO_3$	$\dfrac{d[NO]}{dt} = \dfrac{0.1\,[NO]}{M \cdot s}$	1.0×10^6 sec
8.3	$NO_2^- + H_2O_2 \rightarrow NO_3^- + H_2O$	$\dfrac{d[H_2O_2]}{dt} = \dfrac{1.38 \times 10^2\,[H^+]\,[HNO_2]\,[H_2O_2]}{M^2 \cdot s}$	1.6×10^{10} sec
8.4	$NO_2^- + Cl_2 + H_2O \rightarrow NO_3^- + 2\,H^+ + 2\,Cl^-$	$\dfrac{d[Cl_2]}{dt} = (k_3 + k_4[HNO_2])\dfrac{[Cl_2]\,[NO_2^-]}{[Cl^-]^2}$	(valid for pH 0–1)
8.5	$N_2O_3 + H_2O \rightarrow 2\,HNO_2$	$\dfrac{d[N_2O_3]}{dt} = \dfrac{5.3 \times 10^2\,[N_2O_3]}{s}$	7.5×10^5 sec
8.6	$N_2O_4 + H_2O \rightarrow HNO_2 + H^+ + NO_3^-$	$\dfrac{d[N_2O_4]}{dt} = \dfrac{1 \times 10^3\,[N_2O_4]}{s}$	8.52×10^4 sec

dominant at acidic pH. Oxygen reactions are disfavored by the low trace metal catalyst concentrations and the low buffering of basic gases, as shown by the low partial pressure of ammonia in equilibrium with the rainwater, as given in Table 3. Measurements of aqueous ozone and hydrogen peroxide concentrations are needed to resolve the relative importance of their oxidation reactions.

Assuming the same scavenging ratio for aerosol sulfate as that for aerosol nitrate, the scavenging of 6.6 $\mu g/m^3$ sulfate aerosol (median value for Pasadena during precipitation) implies a rainwater concentration of roughly 26 μN for sulfate. About two-thirds of the average sulfate in Pasadena precipitation could be a result of aerosol scavenging. Thus aerosol scavenging appears to be important in rainwater scavenging of sulfates, and the remainder of the sulfur species in precipitation may be a result of pH-controlled equilibrium dissolution of SO_2 and subsequent oxidation in the raindrop.

A comparison of the sulfate and nitrate wet deposition fluxes with the source emission fluxes into the air gives an estimate of the overall oxidation and scavenging efficiency of these air pollutants. Approximately 40% of the hourly sulfur emissions are scavenged during precipitation. In contrast, 13% of the hourly NO_x emissions are scavenged during storms. These results concur with those found by Hales and Dana [52] for the St. Louis METRO-MEX region.

REFERENCES

1. Oden, S. Naturvetenskaphga Forkingsrad, Ekologokommitteen, Stockholm, Bull. No. 1 (1968).
2. Cogbill, C. V., and G. E. Likens. *Water Resources Res.* 10:1133 (1974).
3. Granat, L. *Tellus* 24:550 (1972).
4. Knudsen, E. J., et al. *ACS Symp. Series* 52:80 (1977).
5. Dana, M. T., J. M. Hales and M. A. Wolf. *J. Geophys. Res.* 80:4119 (1975).
6. Kennedy, V. C., G. W. Zellweger and R. J. Avanzino. *Water Resources Res.* 15:687 (1979).
7. Liljestrand, H. M., and J. J. Morgan. *Environ. Sci. Technol.* 12:1271 (1978).
8. McColl, J. G., and D. S. Bush. *J. Environ. Qual.* 7:352 (1978).
9. Lewis, W. M., Jr. and M. C. Grant. *Science* 207:176 (1980).
10. Liljestrand, H. M., and J. J. Morgan. *Environ. Sci. Technol.* 15:333 (1981).
11. Gran, G. *Analyst* 77:661 (1952).
12. Seymour, M. D., et al. *Water Air Soil Poll.* 10:147 (1978).
13. *Standard Methods for the Examination of Water and Wastewater,* APHA, AWWA, WPCF, American Public Health Assoc., Washington, D.C., 1975.

14. Florence, T. M., and Y. J. Farrar. *Anal. Chim. Acta* 54:373 (1971).
15. Strickland, J. D. H., and E. Parsons. "A Practical Handbook of Seawater Analysis," Fisheries Research Board of Canada (1960).
16. Miller, M. S., S. K. Friedlander and G. M. Hidy. *J. Colloid Interface Sci.* 39:165 (1972).
17. Liljestrand, H. M., and J. J. Morgan. In: *Polluted Rain*, T. Y. Toribara, Ed. (New York: Plenum Publishing Corporation, 1980).
18. Slinn, W. G. N., et al. *Atmos. Environ.* 12:2055 (1978).
19. "1975 National Emissions Report," U.S. EPA–450/2–78–020 (1978).
20. Cass, G. R. "Methods for Sulfate Air Quality Management with Applications to Los Angeles," PhD Thesis, California Institute of Technology (1978).
21. Sillen, L. G., and A. E. Martell. "Stability Constants of Metal-Ion Complexes," Spec. Pub. 17; Chemical Society, London (1964).
22. Sillen, L. G., and A. E. Martell. "Stability Constants of Metal-Ion Complexes," Spec. Pub. 25; Chemical Society, London (1971).
23. "JANAF Thermochemical Tables," 2nd ed.; U.S. NBS NSRDA–37 (1971).
24. "Selected Values of Chemical Thermodynamic Properties; Tables for the First Thirty-Four Elements in the Standard Order of Arrangement," U.S. NBS, Tech. Note 270–3 (1968).
25. Robie, R. A., B. S. Hemingway and J. R. Fisher. "Thermochemical Properties of Minerals and Related Substances at 298.15°K and 1 Bar (10^5 Pascals) Pressure and at Higher Temperatures," U.S. Geological Survey Bulletin 1452 (1978).
26. Hales, J. M. *Atmos. Environ.* 6:635 (1972).
27. Hales, J. M., and M. T. Dana. *Atmos. Environ.* 13:1121 (1979).
28. Abel E., and E. Neussar. *Monatshefte für Chemia* 54:855 (1929).
29. Abel, E., and H. Schmid. *Z. Physikal. Chem.* 136:430 (1928).
30. Dawson, G. A. *Atmos. Environ.* 12:1991 (1978).
31. Lau, N. C., and G. A. Charleson. *Atmos. Environ.* 11:475 (1977).
32. Hales, J. M., and D. R. Drewes. *Atmos. Environ.* 13:1133 (1979).
33. Kok, G. L. *Atmos. Environ.* 14:653 (1980).
34. Beilke, S., and H. W. Georgii. *Tellus* 20:435 (1968).
35. Froessling, N. *Gerlands Beitr. Geophys.* 52:170 (1938).
36. Dana, M. T., and J. M. Hales. *Atmos. Environ.* 10:45 (1976).
37. Raynor, G. S. "Meteorological and Chemical Relationships from Sequential Precipitation Samples," BNL 22879 Brookhaven National Lab, Upton NY (1977).
38. Georgii, H., and H. Weber. "The Composition of Individual Raindrops," Technical Note AF 61(052)–249 Air Force Cambridge Research Center, Bedford, MA (1960).
39. Wolaver, T. G., and H. Leith. *Edv. Medizin Biol (Stuttgart)* 4:74 (1973).
40. Pearson, F. J., and D. W. Fisher. U.S. Geol. Surv. Water Supply Paper 1535–P (1971).
41. Lodge, J. P., Jr., et al. "Chemistry of United States Precipitation," National Center for Atmospheric Research, Boulder, CO (1968).
42. Vermeulen, A. J. *Environ. Sci. Technol.* 12:1016 (1978).
43. Lazrus, A. L., E. Lorange and J. P. Lodge, Jr. *Environ. Sci. Technol.* 4:55 (1970).

44. Ralston, A., and H. S. Wilf. *Mathematical Methods for Digital Computers* (New York: John Wiley & Sons, 1960).
45. Pogrebnaya, V. L., et al. *Zh. Prikl. Khim.* 48:954 (1975).
46. Atroshchewko, V. I., A. I. Tseitlin and A. V. Shapka. *Zh. Prikl. Khim.* 48:954 (1975).
47. Halfpenny, E., and P. L. Robinson. *J. Chem. Soc.* (1952), p. 928.
48. Pendlebury, J. N., and R. H. Smith. *Aus. J. Chem.* 26:1847, 1857 (1973).
49. Gratzel, M., et al. *Ber. Bunsenges.* 73:646 (1969).
50. Gratzel, M., S. Taniguchi and A. Henglein. *Ber. Bunsenges.* 74:488 (1970).
51. Orel, A. E., and J. N. Seinfeld. *Environ. Sci. Technol.* 11:1000 (1977).
52. Hales, J. M., and M. T. Dana. *J. Appl. Meteorol.* 18:294 (1979).
53. Scott, B. C. *J. Appl. Meteorol.* 17:1375 (1978).
54. Miller, D. F., et al. *Science* 202:1186 (1978).
55. Falls, A. H., G. J. McRae and J. N. Seinfeld. *Int. J. Chem. Kin.* 11:1137 (1979).
56. Penkett, S. A., et al. *J. Atmos. Environ.* 13:123 (1979).

CHAPTER 8

ATMOSPHERIC DEPOSITION OF NITROGEN
AND SULFUR IN NORTHERN CALIFORNIA

John G. McColl

Department of Plant and Soil Biology
University of California
Berkeley, California 94720

INTRODUCTION

Atmospheric precipitation is comprised of dry components (particulates and gases) and wet components (rain, snow, fog). Ocean salts, natural airborne dust, soil particles, gaseous air pollutants, and the rate, frequency and distribution of rainfall, all contribute to chemical characteristics of these components.

The general purpose of the study reported in this chapter was to monitor the main inorganic constituents of atmospheric deposition in northern California, and to determine whether California has an "acid rain" problem. It is important, therefore, to understand what acid rain is, how it is formed, and what problems might arise if acid rain occurs.

Acidity of a solution can be described by its pH. The pH scale ranges from 0 to 14, and pH 7 is usually considered to chemically neutral. However, rain is not pure water, but contains various dissolved salts and other substances. Rain in its theoretically normal, unpolluted state is slightly acid, with pH 5.6, this being due to carbonic acid (H_2CO_3) which is a product of the dissolution of atmospheric carbon dioxide (CO_2). Thus, "acid rain" is defined as rain with a pH < 5.6.

The oxides of sulfur and nitrogen (SO_x and NO_x) are oxidized in the atmosphere and form sulfuric and nitric acids (H_2SO_4 and HNO_3), which contribute to the acidity of rain. Chlorine emissions may also result in acidity (hydrochloric acid, HCl), but usually only close to emission sources; whereas sulfur compounds (and maybe those of nitrogen) can be transported several hundred kilometers per day in the atmosphere.

Sulfur oxides are primarily emitted from stationary sources such as utility and industrial coal-burning boilers. Nitrogen oxides are emitted from both stationary and mobile sources, especially automobiles. Recent data of the U.S. Environmental Protection Agency [1] show that $\sim 56\%$ of NO_x emitted in 1977 was caused from the burning of fossil fuels by stationary sources, and 40% came from transportation-related sources. Emissions of NO_x from stationary sources are likely to increase rapidly in the next 20 years, as combustion of fossil fuels is expected to rapidly increase in this period also [1].

The combustion of fossil fuels in the United States results in about 50 million metric tons of SO_x and NO_x being emitted to the atmosphere. In 1977, SO_x accounted for 14% (27.4 million metric tons) of the total air pollution and NO_x accounted for 12% (23 million metric tons) [1]. These huge emissions of air pollutants result in acid rains, which are now common phenomena in the northeastern United States [2], and also in Scandinavia, where some of the earliest work in this area was performed [3]. The acidity of rain has been increasing in the northeastern United States [2, 4, 5] and is having adverse ecological effects, such as degradation of water quality, fish productivity and possibly forest productivity, and may also cause accelerated soil leaching. Similar effects have been widely documented by Scandinavian workers who have taken pioneering roles in studying acid rain effects and in monitoring acid deposition [6].

The monitoring of acid rain at various locations in the world and results of some studies on its effects, were documented at the First International Symposium on Acid Precipitation and the Forest Ecosystem [7, 8]. The following reports also are useful general references on the subject, and essentially represent the present "state-of-the-art": References 1, 2, 5, and 9-15. At the recent international conference on "The Ecological Impact of Acid Precipitation," sponsored by the Norwegian SNSF-project and held at Sandefjord, Norway in March 1980, ecological effects of "acid rain" were discussed in detail [16].

There has been much less concern about atmospheric deposition in the western United States than in Scandinavia or the northeastern United States, probably because there are few data available for the western United States that either document the chemical characteristics of precipitation, or document environmental degradation caused by changes in precipitation chemistry [17-20]. Also, emissions of SO_x and NO_x are not as great in the West as in the eastern United States, and thus ecological effects from acid rain are thought to be less. For example, in the San Francisco Bay area, a 5-year average concentration of airborne NO_3^- was 2.78 $\mu g/m^3$ (the national urban average was 2.40 $\mu g/m^3$), and the 5-year average for SO_4^{2-} was 2.68 $\mu g/m^3$, which is only slightly above the remote nonurban background level of 2.51 $\mu g/m^3$ [21].

Data on the chemistry of atmospheric precipitation in the western United States are fragmentary, and even fewer data exist on the acidity of wet precipitation in particular. For example, various inorganic ions measured in rain in the United States (including the western states) were reported by Junge and Werby [22] but no pH measurements were made. Whitehead and Feth [23] of the U.S. Geological Survey (USGS) monitored precipitation chemistry at Menlo Park on San Francisco Bay, California, in 1957-1959. The USGS has subsequently reported on rain at Menlo Park for a short period in 1971 [24], and for a three-month period in 1971-1972, at both Menlo Park and Petrolia [17] which is located near the California coast about 500 km north of San Francisco. At Menlo Park, they found that the pH of rain averaged 5.9 in 1957-1958, 5.3 in 1978-1979, and ranged from ~4.5 to 6.0 in 1971. Liljestrand and Morgan [25] monitored rain at Pasadena, where the mean pH was 4.06 in 1976-1977, with nitric acid being 32% more important to the acidity than was sulfuric acid. These workers extended their study to nine locations in southern California in 1978-1979 [19]. Acid precipitation also has been measured at Richmond and Livermore in California by the Department of Energy [26].

Acid rain has been recorded at Boonville and Sacramento, California [27]; the lowest pH at Boonville was 3.6, and at Sacramento was 3.5 during the two rainy seasons of 1977-1978 and 1978-1979. Acid rain also was measured at Lake Tahoe and Davis in 1971 [28]. At Hopland, California (at the University of California Field Station), sulfur and nitrogen depositions have been measured in rain since 1958, but not until this present study has pH been seriously measured there [29].

Similarly, acid rain has been measured during a short period in the Seattle-Tacoma area of Washington state, at distances from the major SO_2 sources at the Tacoma Smelter and nearby refineries [30]. In other locations in the west, various inorganic constituents of rain have been monitored for different purposes, but these generally have not included measurement of pH (e.g. in Utah [31], in southern California [32]).

The author and associates have measured precipitation chemistry (including pH) at Berkeley, California, on an event basis since 1974, except during the drought of 1975-1977. The average pH was 5.0 in 1974-1975 [33], 4.8 in 1977-1978 [34] and 4.7 in 1978-1979 [20].

Obviously, "acid rain" is a common phenomenon in the San Francisco Bay area and other locations in California. Research now must be directed toward a fuller description of its occurrence, and on its possible effects, especially those on the water, plant and soil resources.

The overall objective of the general research project was to monitor both wet and dry components of atmospheric precipitation, to determine the geographical extent and temporal variations of "acid rain" in northern California, and to anticipate some possible ecological effects. In this report, depositions of nitrogen and sulfur and relationships with acidity of wet precipitation are considered in particular. Further details of the general study, which included analyses of 13 different inorganic ions, are reported elsewhere [20].

METHODS

Eight sampling sites were located in northern California: at Berkeley, Tahoe City, Kearney (Parlier), Challenge, San Jose, Hopland, Davis and Napa (Figure 1). Other details of the sites are given in Table 1.

The sites were chosen as representative of different geographical and/or vegetation and land-uses. They also included pollution "source-areas" and potential "receptor areas" where ecological effects may be important, and they represent a small network, each site being close enough to each other for interpretative purposes, e.g., to infer possible atmospheric transport between "source" sites (primarily Berkeley and San Jose) and "receptor sites" (all others).

Separate samples of both wet precipitation and dry fallout were collected using a two-bucket system with a movable lid, designed to expose the wet bucket and cover the dry bucket during periods of wet precipitation, and vice versa (Figure 2). This instrument (Aerochem Metrics Model 201) is one of the few recommended by Galloway and Likens [35], and also by a subcommittee that established guidelines for precipitation measurements in the U.S. National Atmospheric Deposition Program. A sensor, mounted on the frame, reacts electrically to the onset of precipitation causing the lid to move. Heaters are mounted below the sensor to melt both snow and ice and to evaporate moisture from the sensing element.

Sampling and subsequent chemical analysis were made after each main storm event at each site. The pH of each sample was measured in the field at each of the 8 sites, and in the central laboratory using a Corning Digital 112 Research Meter, calibrated with pH 4 and pH 7 buffers using the lower buffer first. Twenty-five mL of solution was poured into a plastic beaker, and the meter was allowed to equilibrate for 5 minutes before the pH reading was recorded. Laboratory pH measurements are reported here.

Figure 1. Sites of sampling atmospheric precipitation in northern California, 1978–1979.

After filtration with a 0.45-μ Millipore filter and chemical digestion with boiling nitric acid plus hydrogen peroxide, cations were determined by atomic absorption spectrophotometry using a Varian, Model AA6 [36]. Chloride was determined using a spectrophotometric method [37], sulfate

Table 1. Description of Study Sites and Identification of Field Collectors

Site Designation	Location and Elevation	Description	Distance from Ocean (km)	Field Collector's Name and Institution
BE	Berkeley, 37°53′, 122°15′, 400 m	Space Science Laboratory, University of California; industrial and urban; pollution-source area; oceanic influence.	20	L. Monette and D. Bush, University of California, Berkeley
TC	Tahoe City, 39°08′, 120°10′, 2,076 m	Sierra forest; watershed and recreation area.	260	R. Leonard, University of California Davis, and Tahoe Research Group
KE	Kearney, 36°46′, 119°43′, 100 m	Horticultural Field Station at Parlier San Joaquin Valley; agriculture.	170	R. Brewer, University of California Davis, at Kearney Agricultural Field Station, Parlier
CH	Challenge, 39°39′, 121°21′, 790 m	Ranger Station, USDA Forest Service, Expt. Station; Mixed conifer forest, lower Sierran foothills.	215	M. Heath, USDA Forest Service
SJ	San Jose, 37°21′, 121°54′, 22 m	San Jose State University; industrial and urban; pollution-source area; oceanic influence.	30	Jindra Goodman and Susan Fisher, San Jose State University
HO	Hopland, 39°00′, 123°03′, 165 m	University of California Field Station, Coast Range; grazing and watershed.	40	M. Jones, University of California Davis, at Hopland Field Station
DA	Davis, 38°32′, 121°46′, 18 m	University of California Davis campus field plot; agricultural and urban.	100	G. Malyj, University of California Davis, and Tahoe Research Group
NA	Napa, 38°17′, 122°16′, 280 m	Woodland on the ridge above Wooden Valley, Coast Range; woodland and agricultural.	55	M. Linn, John Muir Institute

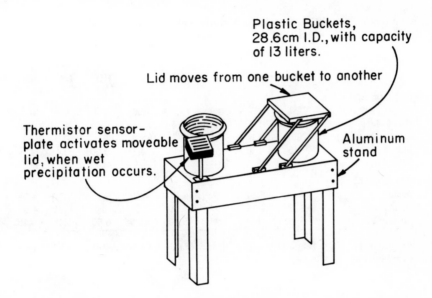

Figure 2. Collector used for event monitoring of wet and dry atmospheric precipitation.

using a barium chloroanilate spectrophotometric method [38], nitrate by a hydrazine reduction method, and ammonium by the indophenol method [39].

All data were punched on computer cards and also recorded on disks at the Computer Center, University of California, Berkeley (UCB). Data were summarized and subsequent statistical analyses were made using the SPSS package programs (i.e., "Statistical Programs for the Social Sciences" [40]), which are on file at the UCB Computer Center.

RESULTS

Storms monitored at each site during the wet season November 1978 through May 1979, numbered as follows: Berkeley, 25; Tahoe City, 35; Kearney, 23; Challenge, 17; San Jose, 20; Hopland, 24; Davis, 26; and Napa, 17. Wet precipitation occurred as rain, except for 13 events at Tahoe City and 8 events at Challenge that were snow.

Detailed statistics of the acidity of wet precipitation are given in Table 2. "Acid rain" commonly occurred. San Jose was the site with the lowest mean pH of the storms sampled, with a mean of 4.42, and San Jose also recorded the storm having the lowest pH of any site with a value of 3.71. Davis was the site with the highest mean pH of 5.20 (Table 2).

Table 2. Statistics of Acidity of Wet Atmospheric Precipitation During the
Wet Season, November 1978 through May 1979

	Site[a]							
	BE	TC	KE	CH	SJ	HO	DA	NA
$[H^+]$ (μequiv/L)								
Mean	22.1	6.7	10.9	13.0	38.0	7.9	6.3	14.6
SE of Mean	3.8	1.0	3.8	1.5	21.3	1.3	1.1	1.5
Maximum	70.8	28.8	66.1	25.1	195.0	29.5	20.0	26.3
Minimum	6.3	1.3	0.4	5.0	0.5	1.9	0.4	4.0
Number of Samples	25	35	23	17	20	24	26	17
Corresponding pH								
Mean	4.66	5.17	4.96	4.88	4.42	5.10	5.20	4.84
Minimum	4.15	4.54	4.18	4.60	3.71	4.53	4.70	4.58
Maximum	5.20	5.90	6.40	5.30	6.29	6.73	6.45	5.40

[a]BE = Berkeley, TC = Tahoe City, KE = Kearney (Parlier), CH = Challenge, SJ = San Jose, HO = Hopland, DA = Davis, NA = Napa.

Relatively high maximum-pH values were recorded for storms at Kearney (pH 6.40), San Jose (pH 6.29), Hopland (pH 6.73) and Davis (pH 6.45). The high pH at San Jose was probably due to upwind emissions of alkaline particulates from a cement factory to the east [41], and presumably due to dissolved soil particles at the other sites which are representative of range and agricultural areas. The wide range of pH at San Jose (pH 3.71-6.29) illustrates the necessity for event monitoring (in contrast to weekly or monthly monitoring, for example), to provide data for interpretative purposes. Such pH ranges may not have been identified if collection times had been determined at regular intervals rather than being dictated by the timing of wet precipitation events.

Frequency distributions of $[H^+]$, $[NO_3^-]$ and $[SO_4^{2-}]$ in wet precipitation at the study sites are shown in Figures 3A-C, where frequency is the number of individual storms with a given concentration. The Berkeley site has the "flattest" distribution curve for H^+, and Tahoe City site had the lowest $[H^+]$ class (Figure 3A). The Kearney site had the flattest and most widespread distribution for $[NO_3^-]$ (Figure 3B). The $[SO_4^{2-}]$ data were generally more evenly distributed for all of the sites (Figure 3C).

Simple linear correlation coefficients were calculated to determine which anions were most closely related to the H^+ in wet precipitation; these results are shown in Figure 4, where geographical location of sites is also considered. Significant correlations of $[H^+]$ with $[NO_3^-]$ occurred in Berkeley, San Jose, Napa, Davis, Hopland and Challenge, but not at Tahoe City or Kearney. Tahoe City was the only site with $[SO_4^{2-}] > [NO_3^-]$ (Table 3), and with the

Table 3. Mean Ionic Concentrations of Wet Precipitation and Total Wet and Dry
Depositions of H^+, NH_4^+, NO_3^- and SO_4^{2-}, During the Wet Season,
November 1978 through May 1979

Ion	Site[a]							
	BE	TC	KE	CH	SJ	HO	DA	NA
Wet concentrations								
(μequiv/L):								
H^+	22.1	6.8	10.9	13.0	38.0	7.9	6.3	14.6
NH_4^+	8.0	4.1	40.0	11.9	19.1	9.7	35.5	12.1
NO_3^-	13.7	6.7	43.4	19.9	16.4	11.1	22.6	16.4
SO_4^{2-}	10.2	13.3	13.8	8.6	10.0	6.2	19.0	11.7
Wet Depositions								
(kg/ha):								
H^+	0.083	0.052	0.016	0.121	0.027	0.046	0.021	0.082
NH_4^+	0.589	0.373	1.205	1.481	0.672	0.590	2.454	1.375
NO_3^-	3.447	2.224	4.142	7.912	1.466	3.043	4.957	5.724
SO_4^{2-}	1.969	2.341	1.367	3.067	1.226	1.381	2.634	3.840
Dry Depositions[b]								
(kg/ha):								
H^+	0.014	0.001	0.123	0.001	0.008	0.004	0.001	0.003
NH_4^+	0.470	0.070	0.552	0.097	0.328	0.364	0.453	0.157
NO_3^-	2.826	0.470	1.803	4.284	2.937	1.170	1.449	0.877
SO_4^{2-}	1.248	0.972	0.579	0.658	1.134	0.720	2.058	0.145

[a]BE = Berkeley, TC = Tahoe City, KE = Kearney (Parlier), CH = Challenge, SJ = San Jose,
HO = Hopland, DA = Davis, NA = Napa.
[b]Dry depositions during the long, dry summer period would greatly inflate the values
here, which are for periods between storms in the wet season only.

lowest specific conductance of only 3.9 μmho/cm. The precipitation at Tahoe
City, although generally acid, was probably weakly buffered and thus a
correlation with any one anion species did not exist.

The Kearney site was the only one with a significant correlation between
$[SO_4^{2-}]$ and $[H^+]$, and the correlations of $[H^+]$ with $[NO_3^-]$ was not significant (Figure 4). These results indicate that the source of acidity at Kearney
was different from that of the other sites. It is likely that sulfuric acid in rain
at Kearney originated from air pollution from oil fields upwind. Alternatively, local agricultural practices, such as addition of ammonium sulfate soil
fertilizer, could explain the high correlation between $[H^+]$ and $[SO_4^{2-}]$ at
Kearney.

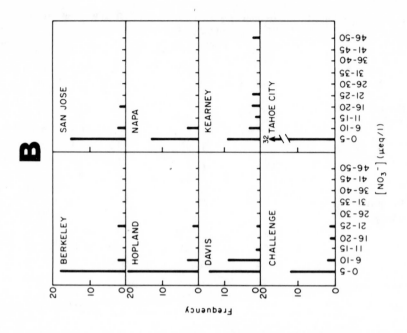

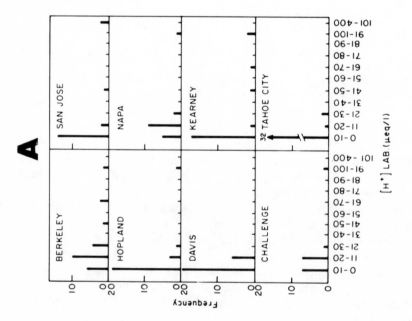

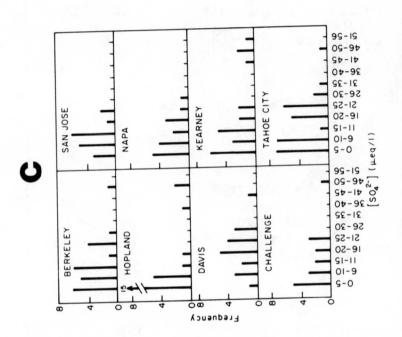

Figure 3. Frequency distribution of ionic concentrations of wet precipitation events at the study sites, 1978–1979: A, $[H^+]$; B, $[NO_3^-]$; C, $[SO_4^{2-}]$.

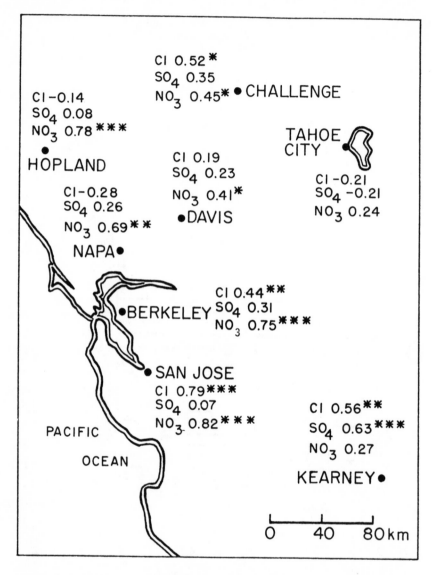

Figure 4. Correlation of $[H^+]$ with anions in wet precipitation, 1978-1979, as indicated by simple linear correlation coefficients.

$$*** \; \alpha < 0.001, \; ** \; \alpha < 0.01, \; * \; \alpha < 0.05.$$

A strong correlation between $[H^+]$ and $[NO_3^-]$ during a large storm in March 1978 also was documented in an earlier study [34] at Berkeley (Figure 5).

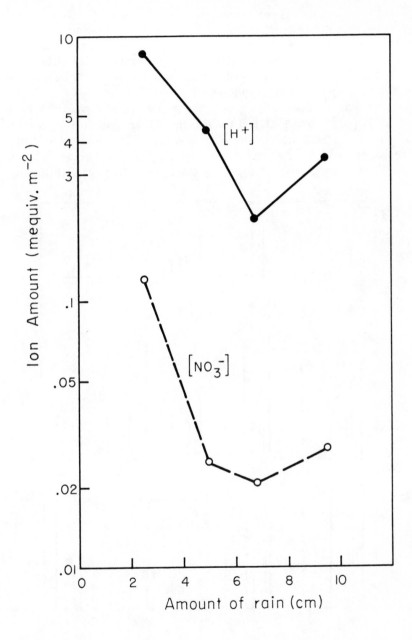

Figure 5. Correlation of [H⁺] and [NO₃⁻] during a three-day rainstorm at Berkeley in March 1978.

These facts strongly suggest that the acidity in the San Francisco Bay area, extending eastward to Challenge via Davis and northward to Hopland via Napa, is largely due to nitric acid. However, significant correlations between $[H^+]$ and $[Cl^-]$ indicate that hydrochloric acid also may contribute to the acidity of wet precipitation (Figure 4).

Ionic concentrations varied widely between storms at a given site. Data of $[H^+]$, $[SO_4^{2-}]$ and $[NO_3^-]$ at the Berkeley site are presented in Figure 6 as examples of the 8 sites monitored. In some cases $[H^+]$ decreased following rainfall, as shown for the series of storms in early January, but this "dilution effect" was not general, as shown by storms in mid-February (Figure 6).

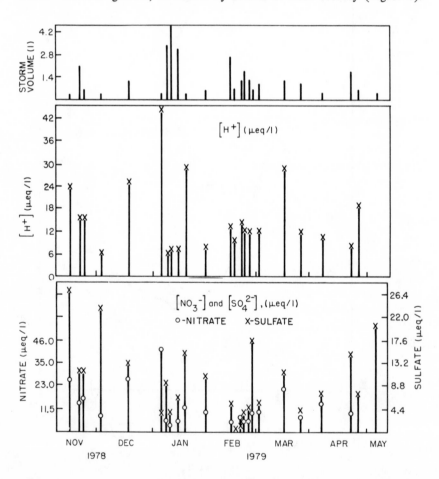

Figure 6. Storm volume; $[H^+]$, $[NO_3^-]$ and $[SO_4^{2-}]$ of wet precipitation at Berkeley, 1978–1979.

The fluctuations in $[H^+]$, $[SO_4^{2-}]$ and $[NO_3^-]$ illustrate the fact that it is difficult to generalize or infer effects from mean data alone. Concentrations of the ions sometimes vary concomitantly, but not always. It is also difficult to generalize about the relationship between storm frequency (or volume) and the corresponding ionic concentrations.

DISCUSSION

As mentioned earlier, only fragmentary data of the chemical composition of precipitation for California exist, but there is some evidence that the pH of rain in the San Francisco Bay area was higher 20 years ago. At Menlo Park, south of San Francisco, rain pH averaged 5.9 in 1957-1958, 5.3 in 1958-1959 [23], and 5.2 in 1971 [24]. At Berkeley, rain pH averaged 5.0 in 1974-1975 [33], 4.8 in 1977-1078 [34], and 4.7 in 1978-1979 (this study). As discussed in an earlier report of the author [33], the decrease in pH of rain seems to be related to $[NO_3^-]$, as shown by the Menlo Park data previously cited: $[SO_4^{2-}]$ dropped from 1.52 ± 0.38 ppm to 0.69 ± 0.18 ppm between 1957 and 1971. During the same period, $[NO_3^-]$ remained almost the same (0.15 ± 0.03 ppm in 1957-1958 and 0.16 ± 0.01 in 1971), even though total ionic concentration of rain water halved. That result is consistent with those of this study, which indicate that primarily nitric acid (and secondarily sulfuric acid) causes the acidity in rainfall in northern California.

Oceanic salts are carried inland and deposited both as dry and wet precipitation as major storm fronts move across California in an easterly direction. For example, Berkeley, which is the site closest to the ocean, received relatively large amounts of both wet and dry depositions of Na^+ and Cl^- [20]. The concomitant fluctuations in $[Na^+]$ and $[Cl^-]$ at Berkeley in relation to storm volume over the study period are shown in Figure 7. Total depositions of ocean salts, however, are also determined by the total amount of rain or snow at a given site as shown by the relative amounts of atmospheric precipitation and the sodium deposition (kg/ha) at the 8 sites shown in Figure 8, relative to their distance from the ocean.

This oceanic contribution to the composition of rain must be distinguished if the contribution from anthropogenic air pollution and/or terrestrial dust are to be quantified also. The fact that most Na^+ or Cl^- in rain is of oceanic origin has been used by various researchers as a basis for determining ionic contributions of other ocean salts to rain water [4, 22, 33, 42].

The mean $[Na^+]$ and $[Cl^-]$ for the eight study sites are plotted in Figure 9 as a function of distance from the ocean. (These distances are simply the distance of the perpendicular drawn from each site to an average line describing the California coast.) Clearly, there is a highly predictable and rather

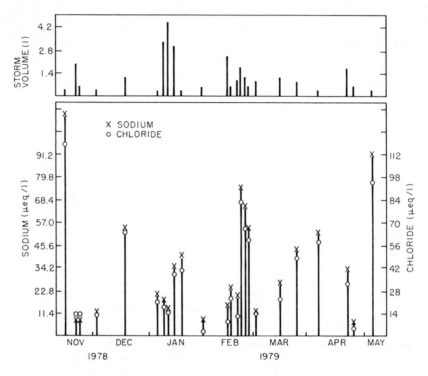

Figure 7. Storm volume, [Na$^+$] and [Cl$^-$] of wet precipitation at Berkeley, 1978–1979.

abrupt decrease in [Na$^+$] and [Cl$^-$], but beyond ~80-100 km inland, the concentrations level off (Figure 9). Beyond ~80 km inland, chloride continues to decrease more than sodium, as soil particles can contribute to some sodium concentrations, whereas the presumed source of chloride is almost all ocean. Thus the ratio of Cl$^-$:Na$^+$ also decreases very predictably with distance from the ocean (Figure 10).

These highly significant functions are useful tools to predict the contribution of other ionic constituents of rain derived from oceanic sources. For example, using [Cl$^-$] as a basis, knowing that the [Cl$^-$] : [NA$^+$] ratio (on a chemical equivalent basis) of ocean water is 1.17 [43] and assuming that all [Cl$^-$] is of oceanic origin, [Na$^+$] can be calculated for all sites if their Cl$^-$is known. Similarly, the concentration of other ions (Ca$^+$, Mg^{2+}, SO$_4^{2-}$, etc.) in rainwater due to oceanic origin can be predicted, as shown in Figure 11. If the measured concentration of any of these ions in rain at each site exceeds that predicted in Figure 11, the excess can be reasonably assumed to be due to a source other than the ocean. Data for nitrogen are not shown in Figure

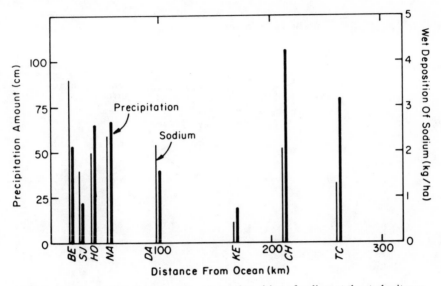

Figure 8. Wet precipitation amount and wet deposition of sodium at the study sites as a function of distance from the ocean, 1978–1979.

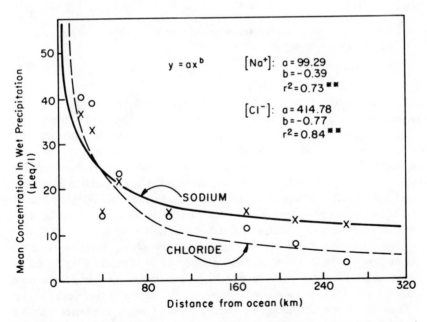

Figure 9. Mean [Na⁺] and [Cl⁻] of wet precipitation at the study sites, as a function of distance from the ocean, 1978–1979.

$X = [Na^+]$, $O = [Cl^-]$, $** = \alpha < 0.01$

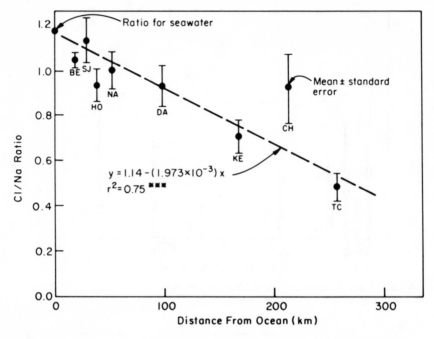

Figure 10. $[Cl^-]:[Na^+]$ ratio of wet precipitation at the study sites, as a function of distance from the ocean, 1978–1979.

*** $\alpha < 0.001$.

11. Because amounts of nitrogen in ocean water are negligible [44], it can be reasonably assumed that all nitrogen in rainwater is of some other origin.

Similar calculations of ion ratios relative to specific elements uniquely derived from known sources allow prediction of quantities of different elements from soil dust or certain other air pollution sources. For example, silica may be a good indicator for soil-derived air and rain pollutants, whereas lead may indicate pollutants from automobile-exhaust emissions. Such indicator techniques have been used to some extent in air-pollution studies [45, 46], but only to a very limited extent in rain-pollution studies.

Nitrogen:sulfur ratios were calculated for atmospheric precipitation at the eight sites (Table 4). For ionic concentration of wet samples, $[SO_4^{2-}]$ values were also "corrected" for ocean-salt contributions; no corrections were made for $[NO_3^-]$. In most cases shown by the results of these calculations (Table 4), amounts of nitrogen exceeded those of sulfur at all sites with few exceptions, e.g., Tahoe City. Similarly, Morgan and Liljestrand [19] found that the NO_3^-/SO_4^{2-} equivalent ratio in rain exceeded 1.0 at four of their nine monitoring sites in southern California.

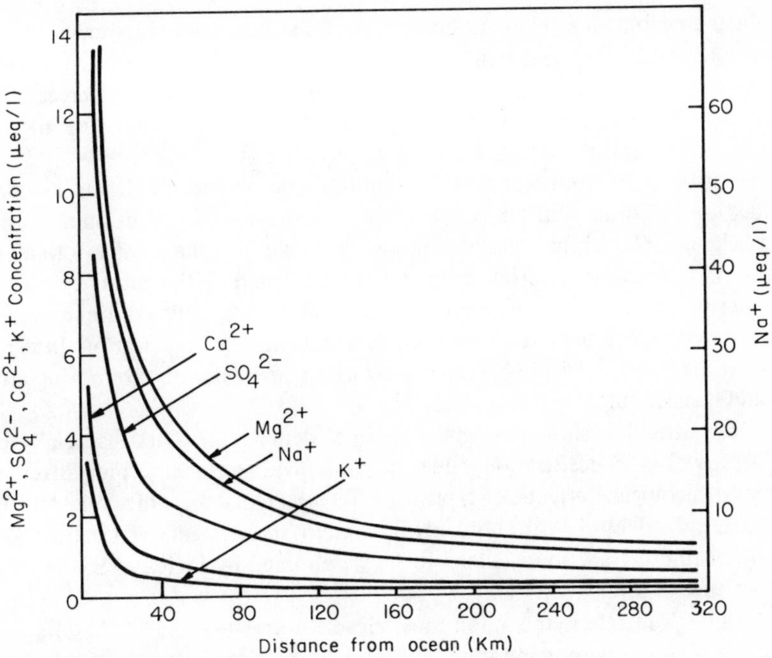

Figure 11. Predicted contribution by oceanic salts to ionic concentrations of wet precipitation as a function of distance from the ocean, 1978-1979. (Based on ratio $[Cl^-]$ in ocean water to $[Cl^-]$ in wet atmospheric precipitation.)

Table 4. Nitrogen:Sulfur Ratios for Atmospheric Precipitation During the Wet Season, November 1978 through May 1979

	Nitrogen:Sulfur Ratios[a]					
Site	1	2	3	4	5	6
Berkeley	1.34	2.36	1.75	1.92	2.26	2.44
Tahoe City	0.50	0.53	0.95	1.03	0.48	0.50
Kearney	3.14	3.36	3.03	4.15	3.11	4.86
Challenge	2.31	2.52	2.58	2.91	6.51	5.39
San Jose	1.64	2.38	1.20	2.84	2.59	2.74
Hopland	1.79	2.85	2.20	2.46	1.63	2.61
Davis	1.19	1.27	1.88	3.48	0.70	2.60
Napa	1.40	1.69	1.49	1.87	6.05	7.36

[a]1 = Wet concentrations, $[NO_3^-]:[SO_4^{2-}]$.
2 = Wet concentrations, $[NO_3^-]:[SO_4^{2-}]$, corrected for ocean-salts.
3 = Wet deposition, $NO_3^-:SO_4^{2-}$.
4 = Wet deposition, Total-N:Total-S.
5 = Dry deposition, $NO_3^-:SO_4^{2-}$.
6 = Dry deposition, Total-N, Total-S.

In the northeastern United States, sulfate is usually the predominant anion associated with the "acid rain" phenomenon. Presumably, emissions of NO_x from automobile exhausts predominate in northern California, whereas SO_x pollution from coal-burning electric-power plants is the main cause of the acidity in the northeastern United States.

Monitoring of both wet and dry atmospheric deposition should be continued in California with objectives similar to those of air-pollution monitoring schemes. A carefully designed network would give the most informative data, and monitoring on an event basis would permit the most legitimate predictions to be made of possible ecological effects. This continued monitoring task would be best undertaken by a state agency organization. Integration with a national United States network of precipitation monitoring also would be advisable.

Close attention should be given to total deposition values (especially of H^+) as well as concentration values, to anticipate and/or interpret different types of ecological effects, e.g., plant or fish sensitivity to acidity may largely be determined by a "threshold pH," whereas accelerated soil leaching and rock weathering are more likely to be determined by increases in the total deposition of acidic material.

Although nitrate is the anion most closely correlated to acidity, sulfate in both wet and dry precipitation also should be given close attention, particularly as sulfur pollution is the primary cause of major acid rain problems elsewhere (e.g., northeastern United States), and because increased sulfur pollution may be anticipated from recently proposed, coal-burning power plants that may be built in California, and from oil-recovery activities.

More research is also needed on the collection procedures and data interpretation of dry fallout, as this type of atmospheric deposition may be even more important than is wet deposition in many areas, especially in California where the wet season is limited to a few months and where the long summers are dry.

ACKNOWLEDGMENTS

I thank Dr. Larry Monette (Post-doctoral Research Associate in my laboratory, University of California, Berkeley), who acted as Coordinator of the project. Special thanks are due to the field operators (mentioned by name in Table 1). Dr. Monette and my laboratory assistants also conducted the chemical analyses. Thanks are also due to Dr. Robert Leonard (of the Lake Tahoe Research Group, University of California, Davis), who did some of the sulfur and all of the nitrogen analyses. Mr. Douglas Bush (a graduate student in my laboratory) helped in many ways, but particularly with statistical analyses and computations. Financial support was provided by the California Air Resources Board, Contract #A7-149-30, and the Agricultural Experiment Station of the University of California, Berkeley. The work was performed at the University of California, Department of Plant and Soil Biology, under Hatch Project #CA-B-SPN-3664-H.

REFERENCES

1. "Acid Rain Research Summary," U.S. EPA Office of Research and Development, U.S. EPA-600/8-79-028 (October 1979), 23 pp.
2. Likens, G. E. "Acid Precipitation," *Chem. Eng. News* 54:29-44 (1976).
3. Barrett, E., and G. Brodin. "The Acidity of Scandinavian Precipitation," *Tellus* 7:251-257 (1955).
4. Cogbill, C. V., and G. E. Likens. "Acid Precipitation in the Northeastern United States," *Water Resources Res.* 10:1133-1137 (1974).
5. Likens, G. E., et al. "Acid Rain," *Scientific Am.* 24L:43-51 (1979).
6. Tollan, A., and L. N. Overrein, Eds. "Annotated Bibliography 1974-1977, SNSF project; Acid Precipitation-Effects on Forest and Fish," (Oslo, Norway: SNSF, 1978), 39 pp.
7. Dochinger, L. S., and T. A. Seliga, Eds. *Proceedings of the First International Symposium on Acid Precipitation and the Forest Ecosystem,* Northeastern, Forestry Experimental Station, Upper Darby, PA. USDA Forest Service Gen. Tech. Rep. NE-23. (1976), 1074 pp.
8. McCormac, B., Ed. *Proceedings of the First International Symposium of Acid Precipitation and the Forest Ecosystem* (1976) and *Water Air Soil Poll.* 6:135-514; 7:279-550; 8:31-129 (1976, 1977).
9. Glass, N. R., G. E. Likens and L. S. Dochinger. "The Ecological Effects of Atmospheric Deposition," EPA, ORD Decision Series, Energy/Environment III, U.S. EPA-600/9-78-022 (October 1978), pp. 113-119.
10. Galloway, J. N., and E. B. Cowling. "The Effects of Precipitation on Aquatic and Terrestrial Ecosystems: A Proposed Precipitation Chemistry Network," *Air Poll. Control Assoc.* 28:229-235 (1978).
11. Hutchinson, T. C., and M. Havas Eds. "Effects of Acid Precipitation on Terrestrial Ecosystems," (New York: Plenum Press, 1980), 654 pp.
12. "Research and Monitoring of Precipitation Chemistry in the United States—Present Status and Future Needs," U.S. Department of the Interior and U.S. Geological Survey Interagency Advisory Committee on Water Data, Office of Water Data Coordination, Reston, VA (1978), 64 pp.
13. "A National Program for Assessing the Problems of Atmospheric Deposition (Acid Rain)—A Report to the Council on Environmental Quality," U.S. National Atmospheric Deposition Program, Colorado State University, Fort Collins, CO (1978), 97 pp.
14. "Ecological Effects of Acid Precipitation," Electric Power Research Institute Report on Workshops at Cally Hotel, Gatehouse-of-Fleet, Galloway, U.K., 4-7 Sept. 1978 (1979).

15. Niemann, B. L., et al. "An Integrated Monitoring Network for Acid Deposition: A Proposed Strategy," Interim Report, U.S. EPA R–023–EPA–79, Teknekron Research Inc., Berkeley, CA. (1979), 235 pp.

16. Drablos, D., and A. Tollan Eds. *Ecological Impact of Acid Precipitation—Proceedings of the International Conference, Sandefjord, Norway, March 11–14,* (Oslo, Norway, SNSF, 1980), 383 pp.

17. Kennedy, V. C., G. W. Zellweger and R. J. Avanzino. "Variation of Rain Chemistry During Storms at Two Sites in Northern California," *Water Resources Res.* 15:687–702 (1979).

18. Lewis, W. M. Jr., and M. C. Grant. "Acid Precipitation in the Western United States," *Science* 207:176–177 (1980).

19. Morgan, J. J., and H. M. Liljestrand. "Measurement and Interpretation of Acid Rainfall in the Los Angeles Basin," Final Report to California Air Resources Board (February 29, 1980), 54 pp.

20. McColl, J. G. "A Survey of Acid Precipitation in Northern California," University of California, Berkeley, Final report to California Air Resources Board (1980), 94 pp.

21. Sandberg, J. S., et al. "Sulfate and Nitrate Particulates as Related to SO_2 and NO_2 Gases and Emissions," *J. Air Poll. Control Assoc.* 26: 559–564 (1976).

22. Junge, C. E., and W. T. Werby. "The Concentration of Chloride, Sodium, Potassium, Calcium and Sulfate in Rainwater over the U.S.," *J. Meteorol.* 15:417–425 (1958).

23. Whitehead, H. C., and J. H. Feth. "Chemical Composition of Rain, Dry Fallout and Bulk Precipitation at Menlo Park, CA., 1957–1959, *J. Geophys. Res.* 69:3319–3333 (1964).

24. Kennedy, V., G. W. Zellweger and R. J. Avanzino. "Composition of Selected Rain Samples Collected at Menlo Park, California, in 1971," U.S. Geological Survey Open File Rep. 76–852 (1976), 7 pp.

25. Liljestrand, H. M., and J. J. Morgan. "Chemical Composition of Acid Precipitation in Pasadena, CA," *Environ. Sci. Technol.* 12:1271–1273 (1978).

26. "The Chemical Composition of Precipitation and Dry Atmospheric Deposition," Environmental Measurements Laboratory, U.S. Department of Energy Publ. EML–356 (1979), pp. I–251 to I–350.

27. Reynolds, R. L. California Air Resources Board Internal Memorandum (July 9, 1979).

28. Leonard, R. Personal communication (1978).

29. Jones, M. Personal communication (1978).

30. Larson, T. V., et al. "The Influence of a Sulfur Dioxide Point Source on the Rain Chemistry of a Single Storm in the Puget Sound Region," *Water Air Soil Poll.* 4:319–328 (1976).

31. Hart, G. E., A. R. Southard and J. S. Williams. "Influence of Vegetation and Substrate on Streamwater Chemistry in Northern Utah," USDI Office of Water Resources Research Project No. A–007–Utah (1973), 53 pp.

32. Schlesinger, W. H., and M. M. Hasey. "The Nutrient Content of Precipitation, Dry Fallout and Intercepted Aerosols in the Chaparral of Southern California," *Am. Midl. Nat.* 103:114–122 (1980).

33. McColl, J. G., and D. S. Bush. "Precipitation and Throughfall Chemistry in the San Francisco Bay Area," *J. Environ. Qual.* 7:352-357 (1978).

34. Bush, D. S. "Mechanisms Controlling Temporal Variation in Throughfall Chemistry," MS thesis, University of California, Berkeley (1979), 152 pp.

35. Galloway, J. N., and G. E. Likens. "Calibration of Collection Procedures for the Determination of Precipitation Chemistry," in *Proceedings of the First International Conference on Acid Precipitation and the Forest Ecosystem.* USDA Forest Service Gen. Tech. Rep. NE-23 (1976), pp. 137-156.

36. Isaac, R. A., and J. D. Kerber. "Atomic Absorption and Flame Photometry:Techniques and Uses in Soil, Plant and Water Analysis," in *Instrumental Methods for Analysis of Soils and Plant Tissues,* L. M. Walsh, Ed. (Madison, WI: Soil Science Society of America, 1971), pp. 17-39.

37. Florence, T. M., and Y. J. Farrar. "Spectrophotometric Determination of Chloride at the Parts-per-Billion Level by the Mercury (II) Thiocyanate Method," *Anal. Chim. Acta* 54:373-377 (1971).

38. Bertolacini, R. J., and J. E. Barney II. "Ultraviolet Spectrophotometric Determination of Sulfate, Chloride and Fluoride with Chloranilic Acid," *Anal. Chem.* 30:202-205 (1958).

39. Solorzano, L. "Determination of Ammonia in Natural Waters by the Phenolhypochlorite Method," *Limnol. Oceanog.* 14:799-801 (1969).

40. Nie, H. N., et al. *Statistical Package for the Social Sciences,* 2nd ed. (New York: McGraw-Hill Book Company, 1975), 675 pp.

41. Goodman, J. Personal communication (1978).

42. Granat, L. "On the Relation Between pH and the Chemical Composition in Atmospheric Precipitation," *Tellus* 24:550-560 (1972).

43. Martin, D. F. "Coordination Chemistry of the Oceans," in *Equilibrium Concepts in Natural Water Systems,* R. F. Gould Ed., Advances in Chemistry Series No. 67 (Washington, DC: American Chemical Society, 1967), pp. 255-269.

44. McGill, D. A. "Light and Nutrients in the Indian Ocean," in *The Biology of the Indian Ocean,* B. Zeitzschel, Ed. (Berlin: Springer-Verlag, 1973), pp. 53-102.

45. John, W., et al. "Trace Element Concentrations in Aerosols from the San Francisco Bay Area," *Atmos. Environ.* 7:107-117 (1973).

46. Miller, M. S., S. K. Friedlander and G. M. Hidy. "A Chemical Element Balance for the Pasadena Aerosol," *J. Colloid Interface Sci.* 39:165-176 (1972).

CHAPTER 9

ACID PRECIPITATION IMPACT ASSESSMENT IN MINNESOTA DERIVED FROM CURRENT AND HISTORICAL DATA

S. A. Heiskary, M. E. Hora and J. D. Thornton

Minnesota Pollution Control Agency
Roseville, Minnesota 55113

INTRODUCTION

The potential impacts of acid deposition on the aquatic resources of the upper midwestern United States have only recently been acknowledged. Most attention and research has been focused in the northeastern United States and eastern Canada, where dramatic effects have been documented. No documentation of lake acidification currently exists in Minnesota; however, the lakes and streams of the central and northeastern parts of the state lie on the southern edge of the Canadian Shield and are highly susceptible to acidification.

Minnesota exists in the transition zone between alkaline and acid precipitation. This is due in part to the prairie-forest transition that crosses from the southwestern to the northeastern corners of the state, and to long-range transport, since Minnesota is not directly downwind of the Midwest industrial complexes. The best available information demonstrates that the precipitation falling in the forested areas (central and northeast) is highly acidic. The volume-weighted mean pH of rain and snow for this area is 4.6–4.8, with individual events reported as low as 3.6. Currently, the annual deposition of

147

sulfate and nitrate ($\sim$16 and 11 kg/ha/yr, respectively) is great enough to cause impact to sensitive aquatic systems [1].

The 1980 Minnesota Legislature designated the Minnesota Pollution Control Agency (MPCA) as the coordinating agency in the state's effort to address the problem of acid rain. One of the important aspects of this program involved the monitoring of lakes and streams throughout the state to determine the extent of acid-sensitive surface waters in the state. This chapter describes the lake investigations conducted in 1980, summarizes pertinent water quality data as they relate to acid rain, and attempts to relate these current data with existing data available for these lakes.

LAKE SELECTION

Two major sampling efforts were conducted in 1980 to assess the extent of acid-sensitive lakes in Minnesota. Table 1 outlines the number of lakes sampled in each survey and the number of samples sent to the State Health Department for complete analysis.

The selection of lakes for these surveys was based on criteria set down by Glass and Loucks [2] as follows:

1. no stream or spring inlets (i.e. headwater lakes);
2. hard rock basins;
3. shallow, coarse or negligible soils throughout the watershed;
4. absence of any significant adjacent wetlands or peat; and
5. watershed in the adjacent land area of about the same as or less than twice the area of the lake.

These criteria as they were used for our lake selection process are expressed as follows:

1. watersheds characterized by granitic bedrock overlain by shallow non-calcareous soils;
2. lakes lacking significant adjacent wetlands;
3. no stream inlets when possible;
4. lakes with small watersheds relative to lake size;
5. lakes or watersheds with some historical water chemistry data indicating low pH or alkalinities; and
6. lakes accessible from a road.

Table 1. MPCA Acid Rain Lake Sampling Program: 1980

	Summer (May–June)	Fall (October–November)	Total
Total lakes sampled	84	108	192
New lakes sampled	84	90	174
Analyzed by lab	78	103	181

Knowledge of bedrock and soil geology were instrumental in the lake selection process for the summer survey. This information identified portions of the state potentially susceptible to acid precipitation, from a geological standpoint, i.e., granitic bedrock overlain by shallow noncalcareous soils. Historical water chemistry data served to further identify the study areas.

The criteria used to select lakes for the fall survey differed slightly from the summer survey. Our initial plan was to conduct an extensive fly-in survey of the Boundary Waters Canoe Area (BWCA) and the surrounding area. This plan had to be abandoned due to the early formation of ice in 1980. Instead, we opted to conduct an extensive survey of the more accessible lakes in the northeastern portion of the state in an effort to assess the extent of acid-sensitive lakes in this region. We employed many of the previously noted criteria in the selection of the lakes with the omission of historical data and lake size as bases for lake selection. Approximately 15% of the lakes sampled in this survey were repeats of the summer survey.

Both surveys were conducted on lakes accessible from a road. This factor had an important bearing in the lake selection process and led to the exclusion of some of the most sensitive lakes in the state (e.g., BWCA). However, it encouraged us to include lakes of various sizes. Figure 1 compares the lakes included in our surveys to lakes in Minnesota on a surface-area basis.

An understanding of the lake selection process used for each survey is essential to the interpretation of the data that result from the survey. In the case of the summer survey we sought to identify sensitive lakes throughout the state and thus relied heavily on the criteria listed above, to select lakes for this survey. The intent of the fall survey was slightly different. That survey was a concentrated effort to assess the acid sensitivity of a region (i.e., northeastern Minnesota) known to contain numerous sensitive lakes. Thus the data from the fall survey are probably a good reflection of the acid sensitivity of the more accessible lakes of that region.

METHODS: COLLECTION OF SAMPLES AND ANALYSIS

Most of the samples from the summer survey were collected at midlake at a depth of one meter. If canoe or boat access was not possible, wading shoreline grabs were taken. Samples were collected from areas of open water whenever possible to avoid the localized influences of macrophytes on the water chemistry, in particular pH [43]. All fall samples were collected as shoreline grabs, many of which were under ice. Care was taken to avoid contamination by disturbed sediments or melt water on the ice.

Field measurements of pH and alkalinity were conducted daily for each of the lakes sampled, using an Orion 399A pH meter. Prior to each use the

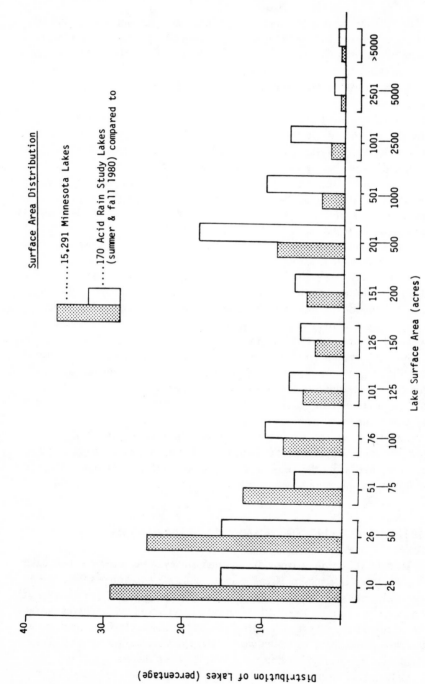

Figure 1. Distribution of lake surface areas for MPCA summer and fall 1980 acid rain study lakes [3].

meter was calibrated with pH 4 and 7 buffers. U.S. Environmental Protection Agency (EPA) Method 150.1 [5] was used to measure pH. Results were reported to the nearest 0.1 pH unit. Alkalinities were measured by the fixed endpoint technique (EPA Method 310.1 [5]). Titrations were carried out with 0.02 N H_2SO_4 and a microburet. The low alkalinity method was used for samples with alkalinities < 20 mg/L as $CaCO_3$. The literature [6] indicates that the fixed endpoint titrations (e.g., pH 4.5) overestimate the actual bicarbonate concentration of dilute waters. Henriksen [7] indicated that actual bicarbonate concentrations are overestimated by 1.6 mg/L as $CaCO_3$ if the fixed endpoint of pH 4.5 is employed.

Gran analysis techniques have been found to be the most accurate method of measuring low alkalinities [8, 9]. In an effort to assess the accuracy of our fixed endpoint titrations, we conducted Gran analysis on a few of our samples for comparison. The mean difference between the fixed-endpoint data and the Gran titration data was 1.1 (mg/L as $CaCO_3$), indicating good agreement between the two methods. For the purpose of this study, the fixed-endpoint data are utilized to maintain consistency.

A quality control check was conducted to assess the precision of our field analyses of pH and alkalinity. Four persons who conducted the field analyses participated in the test. This test was conducted on two EPA quality control samples.

The standard deviations of the pH and alkalinity measurements were less than or equal to those indicated as attainable by *Standard Methods* [10] in terms of precision (reproducibility) of the method. Spot checks in the field also indicated comparable levels of precision for the two techniques. A statistical assessment of the accuracy of our techniques is not advised due to the small number of samples (2). The data, both the individual measurements and means, compare favorably with the true value of each sample and thus suggest that the methodologies yield fairly accurate results for these ranges of alkalinity and pH.

The samples sent to the lab (Minnesota Department of Health–Analytical Services) were analyzed for the following parameters: pH, total alkalinity, total hardness, conductivity, calcium, magnesium, sodium, potassium, sulfate, chloride, nitrogen as nitrite-nitrate, Kjeldahl nitrogen, total phosphorus, aluminum, mercury and color. All analyses were conducted according to *Standard Methods* [10]. The detection limits employed for some of the cations (calcium and magnesium) and anions (bicarbonate and sulfate) were not low enough to accurately assess the concentrations of these parameters for many of the softwater lakes in our study, therefore discussion of these ions is not included. All data have been entered into STORET, the EPA's national water quality data bank.

Table 2. MPCA Acid Rain Lake Data (Summer 1980)

County	Lake I.D. [3]	Lake	Acreage	Field pH	Field Alk (mg/L)	Color (Pt-Co)	Lab Al (µg/L)	Lab Hg (µg/L)	Historical Alk (mg/L)	Historical Year
Becker (5)	3-127	Bass	142	8.3	150.0	—	15	0.14	125.0	1963
	3-163	Lizzie	89	8.1	150.0	—	6	<0.1	No Data	No Data
	3-153	Island	1209	8.0	142.0	—	10	<0.1	No Data	No Data
	3-102	Shell	3377	7.2	142.0	—	5	<0.1	155.0	1962
	3-123	Jones	36	7.1	127.0	—	No Data	—	133.0	1955
Carlton (5)	9-006	Blackhoof	41	8.9	11.9	<5	60	0.1	No Data	No Data
	9-008	Chub	274	7.4	47.0	5	11	0.13	47.0	1951
	9-010	Hay	103	7.0	5.5	40	420	0.19	12.5	1955
	9-016	Sandy	123	6.8	2.5	25	53	0.17	17.5	1959
	9-007	Spring	41	7.9	69.5	<5	9	0.2	69.0	1974
Cass (6)	11-045	Margaret	13	5.3	4.5	—	62	<0.1	10.0	1938
	11-046	Marion	8	6.1	5.3	—	26	<0.1	10.0	1955
	11-047	Mule	64	8.3	71.0	—	No Data	No Data	78.0	1967
	11-055	Pavelgrit	20	6.8	3.5	35	38	<0.1	12.5	1956
	11-043	Roosevelt	1561	8.3	131.0	—	No Data	No Data	125	1946
	11-054	Snowshoe (L. Andrus)	24	7.0	3.0	5	18	<0.1	No Data	No Data
Cook (12)	16-257	Babble	23	7.0	2.0	—	No Data	No Data	9.0	1972
	16-388	Bouder	13	7.7	4.0	20	48	<0.1	10.5	1960
	16-033	Chester	49	6.9	3.5	10	24	0.2	10.0	1953
	16-029	Devilfish	417	6.6	2.0	20	30	<0.1	7.5	1955
	16-023	Esther	87	6.7	3.0	30	45	0.17	8.0	1956
	16-380	Gust	159	8.0	5.0	30	98	0.13	7.5	1960
	16-030	Highlander	29	5.6	<1.0	60	260	0.25	No Data	No Data

ID	Name								
16-381	Jock Mock	23	9.1	10.5	10	21	0.18	10.0	1940
16-198	Leo	101	8.0	5.0	10	22	0.16	5.0	1939
16-337	Mayhew	247	7.1	14.0	10	<5	0.25	15.0	1953
16-202	Squint	18	6.9	6.5	20	26	0.12	11.0	1939
16-343	Surber	7	7.0	16.5	30	17	<0.1	7.0	1974
Crow Wing (3)									
18-206	Papoose	91	5.5	4.8	20	37	0.14	17.5	1966
18-207	Squaw	142	5.3	7.0	5	22	0.12	15.0	1966
18-220	Smokey Hollow	131	6.1	11.5	60	28	0.12	15.0	1966
Itasca (11)									
31-704	Batson	107	8.4	146.0	20	12	0.18	No Data	
31-709	Billo	11	8.8	71.0	–	No Data	–	No Data	
31-403	Bosely	31	5.7	1.0	200	410	0.18	5.0	1952
31-654	Burns	171	8.7	100.0	<5	17	0.11	103	1974
31-663	Forest	29	8.4	102.0	50	13	0.1	No Data	
31-600	Hill	42	6.4	4.0	60	62	0.26	12.5	1947
31-586	Johnson	437	7.1	10.5	5	18	0.13	15.0	1955
31-645	Kremer	64	7.3	6.5	<5	23	0.21	15.0	1949
31-188	Lorraine	31	8.0	130.0	10	6	0.1	1.0	1971
31-664	Ranier	81	8.4	61.0	<5	39	0.1	No Data	
31-646	Surprise	22	7.0	5.0	20	84	0.19	15.0	1949
Lake (8)									
38-024	Crooked	292	8.5	16.0	10	25	0.21	27.5	1958
38-028	Echo	46	7.5	22.0	10	12	0.18	37.0	1937
38-029	Goldeneye	10	7.0	5.0	20	17	0.27	No Data	
38-026	Hare	48	7.5	16.5	30	21	0.21	9.0	1956
38-057	Hogback	44	7.3	16.0	<5	<5	028	15.0	1937
38-269	Homestead	50	8.4	11.0	30	92	0.28	15.0	1961
38-033	Ninemile	339	7.9	16.0	10	29	0.2	20.0	1939
38-058	Scarp	48	7.1	8.5	10	23	0.13	11.0	1971
Murray (4)									
51-040	Bloody	248	7.8	194.0	–	1900	<0.1	No Data	
51-043	Fox	174	8.2	270.0	–	310	<0.1	No Data	

Table 2, continued

County	Lake I.D. [3]	Lake	Acreage	Field			Lab		Historical	
				pH	Alk (mg/L)	Color (pt-co)	Al (µg/L)	Hg (µg/L)	Alk (mg/L)	Year
	51–063	Sarah	1176	8.0	165.0	–	170	<0.1	118.0	1948
	51–046	Shetek	3596	8.0	167.0	–	430	<0.1	121.0	1938
Pine (4)	58–127	Little Bass	18	6.4	2.0	20	20	0.1	10.0	1967
	58–063	Lords	36	5.8	13.0	20	58	0.15	No Data	
	58–074	Johnson	37	6.1	19.5	15	51	0.24	12.0	1957
	58–067	Sturgeon	1456	6.5	37.0	<5	66	0.11	35.0	1938
St. Louis (22)	69–589	Astrid	114	6.2	6.0	60	66	0.22	7.5	1940
	69–190	Big	2049	6.6	18.0	20	43	0.15	5.0	1951
	69–199	Ed Shave	97	6.5	5.5	10	24	0.1	17.0	1975
	69–924	Elk	14	7.2	42.0	30	9	0.25	No Data	
	69–120	Everett	123	6.5	3.5	40	110	0.1	7.5	1966
	69–119	First	19	6.5	2.0	50	200	<0.1	12.5	1961
	69–923	Hobson	64	6.5	7.0	20	17	0.13	No Data	
	69–456	Jeanette	638	6.4	6.5	50	110	0.11	7.2	1966
	69–086	Little Sletten	20	6.9	4.0	40	110	0.15	No Data	
	69–590	Maude	88	5.8	3.0	100	190	0.14	17.0	1976
	69–065	Minister	57	6.9	5.5	70	110	0.1	10.0	1948
	69–457	Nigh	40	6.1	4.5	100	140	0.1	17.0	1974
	69–588	Pauline	60	6.2	6.0	50	53	0.13	7.5	1940
	69–081	Regen Bogen	12	7.0	5.5	–	No Data	–	10.0	1939
	69–016	Sand	32	6.6	23.0	60	18	<0.1	22.5	1955
	69–139	Santa Claus	11	5.7	0	120	150	<0.1	2.5	1964
	69–084	Sletten	32	6.7	4.5	20	38	0.12	No Data	
	69–111	Smith	220	8.1	52.0	10	12	<0.1	No Data	

	ID	Name								
	69–920	Stuart	26	6.9	6.0	40	61	0.24	No Data	
	69–017	Warren	36	9.0	23.0	–	No Data	–	22.5	1955
	69–921	Waymier	36	6.8	5.0	50	73	0.27	No Data	
	69–916	Dollar	11	6.5	9.0	60	59	0.39	No Data	
Washington	82–045	Clear	37	8.5	143.0	–	8	0.12	95.0	1961
(4)	82–033	Mays	42	8.2	97.0	–	19	<0.1	85.0	1961
	82–046	Square	193	8.4	132.0	–	No Data	–	92.0	1926
	82–031	Terrapin	127	8.1	91.0	–	10	0.22	No Data	

Table 3. MPCA Acid Rain Lake Data (Fall 1980)

County	Lake I.D. [3]	Lake	Acreage	Field			Lab		Historical	
				pH	Alk (mg/L)	Color (Pt-Co)	Al (µg/L)	Hg (µg/L)	Alk (mg/L)	Year
Carlton (7)	9-032	Big	566	—	—	10	21	<0.1	17.5	1936
	9-006	Blackhoof	41	7.0	11.0	<5	40	<0.1	No Data	
	9-008	Chub	274	7.9	45.0	<5	5	<0.1	47.0	1951
	0-057	Eagle	410	7.5	—	20	20	<0.1	No Data	
	9-010	Hay	103	6.7	6.2	40	250	0.14	12.5	1955
	9-016	Sand	123	6.6	4.8	30	86	<0.1	17.5	1959
	9-007	Spring	41	7.9	73.0	10	15	<0.1	69.0	1974
Cook (31)	16-049	Trout	277	6.8	12.0	20	35	0.13	17.5	1951
	16-328	Iron	138	6.6	10.0	60	56	0.43	17.5	1957
	16-342	E. Pope	44	6.6	7.5	40	40	<0.1	15.0	1957
	16-343	Surber	10	6.8	9.0	30	30	<0.1	7.0	1974
	16-239	Poplar	950	6.7	8.5	40	35	0.13	No Data	
	16-198	Leo	114	6.9	7.6	20	17	<0.1	5.0	1939
	16-496	Sawbill	944	6.6	9.0	30	40	<0.1	12.5	1936
	16-373	Christine	192	7.2	17.6	50	76	<0.1	18.0	1937
	16-366	Holly	78	7.1	19.2	100	68	<0.1	27.0	1969
	16-360	Caribou	714	7.1	22.0	40	51	<0.1	27.4	1951
	16-252	Pike	850	7.1	18.2	<5	13	<0.1	25.0	1950
	16-228	West Bearskin	522	7.3	16.4	10	20	<0.1	30.0	1955
	16-227	Hungry Jack	486	7.3	16.0	20	7	0.15	18.0	1938
	16-139	Clearwater	1537	7.2	13.6	<5	17	0.11	17.5	1956
	16-202	Squint	18	6.6	9.0	40	56	<0.1	11.0	1939
	16-146	East Bearskin	643	6.9	8.0	40	38	0.12	20.0	1948
	16-247	Birch	266	7.2	15.8	10	14	<0.1	24.0	1969

ID	Lake							
16–247	Birch	266	7.2	15.8	10	14	<0.1	24.0 1969
16–337	Mayhew	247	7.1	15.4	10	11	<0.1	15.0 1953
16–356	Gunflint	4047	7.1	17.0	20	12	<0.1	26.0 1938
16–607	Round	14	6.9	11.5	10	8	<0.1	No Data
16–448	Loon	1197	6.6	11.0	10	9	<0.1	20.0 1936
16–346	Cascade	534	6.8	5.6	30	120	<0.1	No Data
16–380	Gust	159	6.4	5.0	50	170	<0.1	7.5 1960
16–382	Lichen	306	6.6	7.6	50	99	<0.1	16.7 1960
16–454	Crescent	836	6.5	6.0	20	33	<0.1	12.5 1954
16–486	Baker	22	6.6	6.8	60	130	<0.1	13.7 1970
16–143	Devil Track	1873	7.0	13.0	50	45	0.12	22.0 1935
16–182	Ball Club	231	6.5	5.2	30	60	<0.1	21.0 1969
16–156	Two Island	858	6.7	10.2	30	50	<0.1	21.0 1954
16–348	Brule	5024	6.5	6.0	30	46	<0.1	12.5 1954
16–406	Homer	516	6.8	6.6	40	89	<0.1	6.8 1970
38–656	Greenwood	1469	5.9	4.0	400	510	<0.1	7.5 1936
38–666	Slate	354	6.8	13.0	400	290	<0.1	27.5 1961
38–664	Dunnigan	84	5.9	2.0	20	71	<0.1	9.0 1961
38–735	Sand	506	6.6	17.0	350	200	<0.1	27.5 1961
38–651	Kane	108	6.4	3.0	20	35	<0.1	7.5 1951
38–650	Marble	159	7.3	36.0	100	62	<0.1	No Data
38–750	Christianson	158	5.0	1.0	300	440	0.13	26.3 1956
38–751	Thomas	157	7.0	28.0	300	270	<0.1	22.5 1951
38–744	Stewart	264	7.5	50.0	<5	25	<0.1	No Data
38–415	Delay	121	7.0	21.0	20	24	<0.1	No Data
38–393	Dumbell	476	7.0	27.0	10	19	<0.1	No Data
38–256	Divide	69	6.2	2.0	10	20	<0.1	34.0 1959
38–255	Tanner	63	6.4	4.0	20	20	<0.1	15.0 1961
38–269	Homestead	50	6.5	10.0	40	66	<0.1	15.0 1961
38–218	Elixer	18	6.7	25.0	10	32	<0.1	30.0 1960
38–057	Hogback	44	6.7	16.0	<5	11	<0.1	15.0 1937

Lake
(26)

Table 3, continued

County	Lake I.D. [3]	Lake	Acreage	Field pH	Field Alk (mg/L)	Color (Pt-Co)	Lab Al (µg/L)	Lab Hg (µg/L)	Historical Alk (mg/L)	Historical Year
	38–053	Dam Five	92	6.8	40.0	20	11	<0.1	No Data	
	38–024	Crooked	292	6.6	13.0	20	26	<0.1	27.5	1958
	38–026	Hare	48	6.6	16.0	80	110	0.15	9.0	1956
	38–029	Goldeneye	10	6.2	6.0	30	53	<0.1	No Data	
	38–028	Echo	46	6.6	23.0	<5	29	<0.1	37.0	1937
	38–033	Ninemile	339	6.7	15.0	20	38	<0.1	20.0	1939
	38–724	Tofte	134	7.7	72.0	<5	7	0.11	7.2	1976
	38–738	Garden	6427	6.9	11.0	160	120	0.12	20.0	1965
	69–004	White Iron	6427	7.0	13.0	160	140	<0.1	16.1	1938
	38–779	Farm	1328	7.0	10.0	50	55	<0.1	20.0	1965
St. Louis (43)	69–023	Indian	59	7.2	31.0	120	130	<0.1	27.5	1964
	69–143	Wolf	529	6.6	36.0	200	270	0.12	16.1	1938
	69–027	Stone	228	7.1	54.0	10	5	<0.1	26.3	1956
	69–028	Little Stone	153	7.2	25.0	20	22	<0.1	No Data	
	69–111	Smith	220	7.2	50.0	30	12	<0.1	No Data	
	69–132	Barrs	134	6.8	30.0	100	72	<0.1	31.3	1956
	69–128	Briar	89	6.4	20.0	10	23	<0.1	30.0	1957
	69–129	Spring	121	6.9	52.0	40	10	<0.1	No Data	
	69–011	Pequaywan	533	7.3	45.0	40	30	<0.1	45.0	1936
	69–112	Bear	125	7.2	3.0	40	35	<0.1	No Data	
	69–528	Little Long	19	6.4	10.5	<5	16	0.12	No Data	
	69–532	Pioneer	83	7.0	26.0	90	54	<0.1	28.0	1965
	69–412	Comstock	458	7.3	54.0	50	19	<0.1	28.1	1938
	69–546	Schubert	218	6.3	6.4	20	33	<0.1	No Data	
	69–542	Wilson	64	—	—	<5	15	<0.1	No Data	

69-545	Cameron	171	7.4	26.0	20	16	0.14	28.0	1956
69-538	Berg	128	7.2	27.0	30	18	<0.1	No Data	1951
69-544	Dinham	210	7.5	39.0	50	17	<0.1	35.0	No Data
69-642	Elliot	393	7.9	92.0	10	21	0.13	No Data	1965
69-644	Fig	90	5.5	1.0	80	160	<0.1	7.5	No Data
69-234	Mirror	21	7.7	70.0	10	27	<0.1	No Data	1959
69-399	Deepwater	18	6.4	6.8	<5	26	<0.1	15.0	No Data
69-394	Flowage	110	7.7	71.0	<5	10	<0.1	No Data	No Data
69-235	Sunshine	80	7.2	21.0	<5	10	<0.1	No Data	No Data
69-231	Jacobs	88	7.2	36.0	90	48	<0.1	No Data	No Data
69-232	Horseshoe	96	6.7	17.0	40	39	<0.1	17.5	1957
69-230	Schultz	202	7.8	71.0	10	16	<0.1	No Data	No Data
69-237	Cameron	73	7.8	62.0	10	21	<0.1	No Data	No Data
69-490	Pike	508	7.8	58.0	20	110	0.16	No Data	No Data
69-489	Caribou	569	7.6	42.0	30	160	0.1	No Data	No Data
69-217	West	99	7.0	28.0	30	16	<0.1	27.5	1948
	Robinson #4								
69-277	Clear	129	6.3	11.0	10	16	<0.1	No Data	No Data
69-285	Eagle Nest	1926	7.5	37.0	20	<5	<0.1	No Data	No Data
69-278	Armstrong	382	7.6	36.0	10	7	<0.1	No Data	No Data
69-161	Wolf	301	7.4	33.0	40	21	<0.1	No Data	No Data
69-116	Mitchell	270	6.8	15.0	40	26	0.14	22.5	1964
69-117	Johnson	473	6.0	5.5	90	360	<0.1	10.0	1965
69-061	One Pine	369	6.5	15.0	70	49	<0.1	15.0	1965
69-115	Bear Island	2667	6.5	15.0	60	31	<0.1	17.5	1952
69-541	Anne	10	6.6	9.5	—	No Data	—	No Data	No Data
69-637	Central	75	7.5	57.0	—	No Data	—	No Data	No Data
69-646	Murphy	356	7.5	60.0	—	No Data	—	No Data	No Data
69-398	—	14	6.4	6.0		No Data	—	No Data	No Data

RESULTS AND DISCUSSION

The field alkalinities and pH collected during the summer and fall surveys are reported in Tables 2 and 3. The measurement of total alkalinity as $CaCO_3$ serves as an indicator of a substance's (e.g., water or soil) ability to buffer or neutralize acids. In the context of our work it reflects the ability of a lake or stream to neutralize the impacts of acidic precipitation—an ability that is a function of complex interactions within a lake and its watershed. In the case of many of the lakes in northern and northeastern Minnesota the shallow, noncalcareous soils and the granitic bedrock afford little buffering capacity for lakes and rivers in this region.

Figure 2 represents the alkalinity distributions for the two surveys. Based on the work of Kramer (as reported by Glass et al. [11]) lakes with alkalinities of 15 mg/L or less as $CaCO_3$ are considered sensitive, and those between 15-30 mg/L are considered potentially sensitive to the effects of acid precipitation. Based on these levels, 69% of the summer study lakes are either sensitive or potentially sensitive. For the fall survey this percentage rose to 74%. A paired difference test was conducted using the alkalinity data from the 17 lakes sampled during both surveys. This test yielded a standard deviation of 2.5 mg/L. This small deviation allows for the pooling of the summer and fall alkalinity data for the purposes of defining sensitive lakes and mapping median concentrations, as depicted in Figure 3. The relative size of each circle in the figure indicates the approximate number of lakes in each group, and the degree of shading indicates the median alkalinity for that group. Figure 3 reveals that sensitive lakes are not limited to only northeastern Minnesota. Sensitive lakes were identified in eight of the eleven counties.

The pH of a solution is a numerical expression of the acidity of a solution and is closely related to the alkalinity of that solution. Figure 4 depicts the pH/alkalinity relationship for our 1980 study lakes. The shape of this curve is similar to that reported by Glass et al. [11]. This figure also indicates that the combined results of the two surveys yield a total of 42% susceptible (sensitive) and 26% potentially susceptible of the lakes sampled in 1980. Further, 6% of the lakes studied fall into a classificaton referred to by Glass et al. [11] as the "Fishery Mean Danger Threshold." This threshold was derived from the experience of pH-related fish reproductive failure in the La Cloche mountain lakes of Ontario [12]. This pH threshold ($\leqslant 6.0$) represents the approximate pH at which various species of fish (e.g., smallmouth bass and walleye) experienced reproduction problems. As pH declined, reproduction ceased in numerous other species. Figure 5 depicts the distribution of field pH values for the two surveys. This figure indicates that of those lakes sampled in the summer, eight lakes (9%) exhibited pH values $\leqslant 6.0$. Four lakes (4%), sampled in the fall, fell in this range.

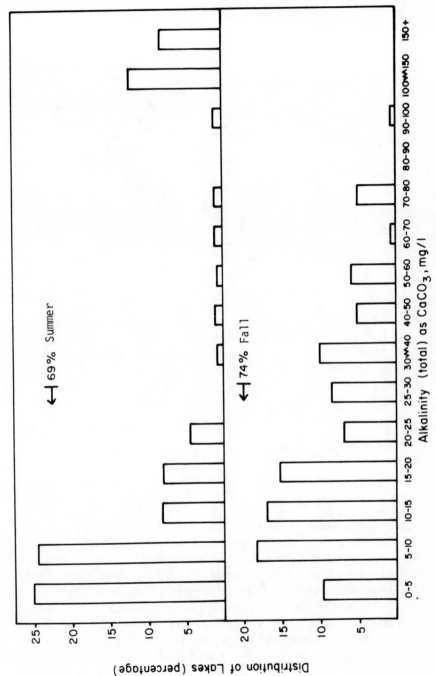

Figure 2. Alkalinity distributions for MPCA 1980 acid rain study lakes.

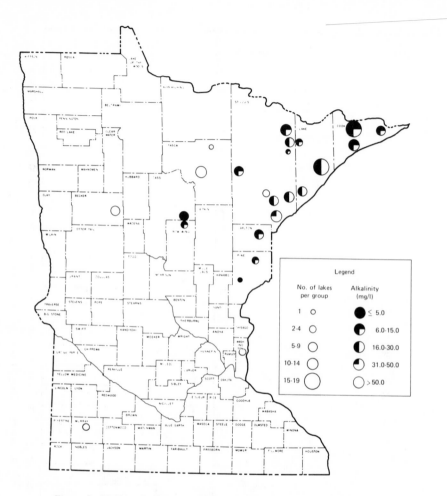

Figure 3. Median alkalinities for MPCA 1980 acid rain study lakes.

Color is an important parameter to consider when attempting to separate the effects of acid precipitation from those of "natural" acidification of a body of water. Figure 6 reveals the color distribution for most of our 1980 study lakes. The relative subdivisions of color are as follows: 0-20 (clear), 21-49 (moderate), and ⩾ 50 (dark), and serve as a comparative scale for the measurement of color status of a water. The numerical scale is the function of a comparative measure against a standardized platinum-cobalt solution [10]. The apparent color in the moderate or dark range is commonly due to dissolved organic matter and in particular humic acids [13]. This coloration is commonly referred to as bog stain. Lakes influenced by a bog "system" are

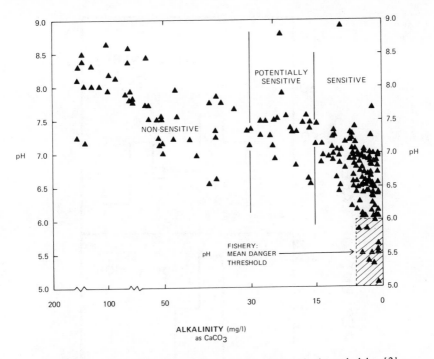

Figure 4. Alkalinity/pH relationship for MPCA 1980 acid rain study lakes [2].

naturally acidic and can yield very low pH values, as low as pH 3 in small bog ponds [14]. The measurement of color allows us to make a preliminary separation of acidification arising from natural processes within the watershed of a lake (i.e., bog influence) from that imparted by atmospheric deposition.

In the case of our 1980 data (Figure 6) 50% of our study lakes could be considered clear and thus probably not affected by bog drainage, and the other 50% most likely have varying degrees of input from bog areas. In reference to the eight lakes with pH values ≤6.0 (summer survey) three are considered clear, and five are considered dark. All four of the lakes with low pH (≤6.0) in the fall survey are considered dark.

This discussion of color as it relates to pH is not meant to suggest that waters with moderate to dark coloration are to be excluded from the monitoring for the effects of acid deposition. Rather it is important to be aware of the acidic inputs which result from the drainage of a bog. The humic components of "bog" water may also give a buffering effect to water over a pH range of 6.0-3.0 [15], and thus these waters may resist acidification better than clear water [16].

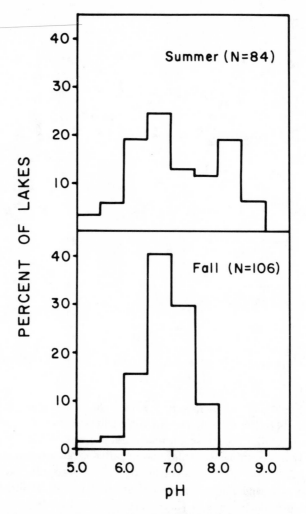

Figure 5. pH distributions for MPCA 1980 acid rain study lakes.

Aluminum is a good indicator of watershed acidification, since precipitation contains only negligible amounts of aluminum [17]. Elevated levels of aluminum (0.2-0.6 mg/L) have been noted for clear acidified lakes in Sweden compared to lakes with normal pH which exhibit levels of 0.05 mg/L or less. These increased concentrations of aluminum also are important from a fisheries standpoint. Cronan and Schofield [18] demonstrated that, in acidified waters, toxic conditions for fish may be produced by dissolved inorganic aluminum even at lake pH values that are not physiologically harmful. Their

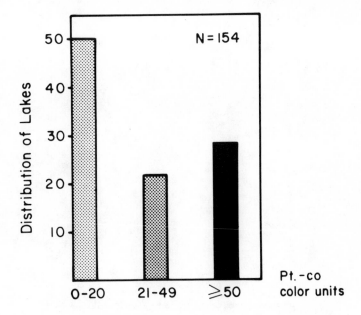

Figure 6. Color distributions for MPCA acid rain study lakes.

studies indicate that brook trout exposed to synthetic acidic solutions and natural Adirondack water with aluminum concentrations >0.2 mg/L showed a specific toxic response to aluminum in the pH range 4.4-5.9. Sublethal reductions in brook trout growth were found in the laboratory at aluminum concentrations of 0.1-0.3 mg/L.

Figure 7 reflects the aluminum-pH (field) relationship for our two surveys. These figures, which include both clear and colored lakes, reveal a weak relationship between aluminum and pH for these lakes. If the colored lakes are ignored this relationship is less apparent. The weakness of this relationship probably reflects the fact that no acidified lakes were identified from our surveys, as reported elsewhere [19]. Also we have very few lakes with pH values from 6.0 to 4.0 (Figure 5), the pH range in which the solubility of aluminum increases dramatically. Four of the lakes from the summer survey lie in the region defined by Cronan and Schofield [18] as having the potential for aluminum toxicity. Two lakes from the fall survey lie in this region.

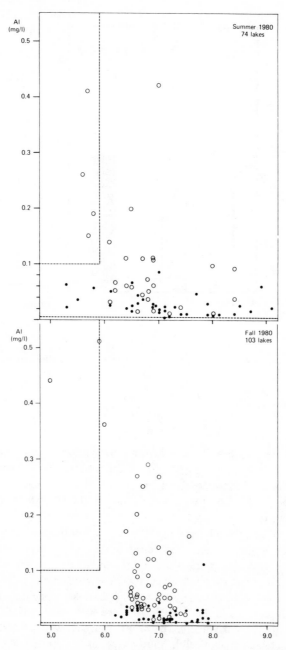

Figure 7. Total aluminum versus pH for MPCA 1980 acid rain study lakes. Closed circles represent clear lakes. Open circles represent colored lakes. Dashed line represents detection limit (0.005 mg/L). Inset rectangle corresponds to aluminum concentrations which may produce toxicity or reduction of growth in fish [18].

Aluminum toxicity is probably not manifested in these lakes for the following reasons:

1. The lakes which exhibit aluminum concentrations in the region of potential toxicity are all either moderate or dark in coloration (>20 Pt–Co units), which implies that these lakes are probably high in humic acids. The literature [20, 21] indicates that these colored waters contain a number of complexing agents which chelate the available aluminum. Apparently this organically chelated aluminum is nontoxic [22].
2. The aluminum measured in our samples is total aluminum. Although this parameter includes both the inorganic and organic fractions, it is an adequate measure for monitoring purposes. Total and inorganic aluminum fluctuate considerably on a seasonal basis and both seem to fluctuate inversely with changes in pH [22]. Organically bound aluminum does not appear to fluctuate with pH but instead tends to correlate well with total organic carbon (TOC) concentrations [21]. This suggests the possible need to monitor TOC in addition to color so that the speciation and toxicity of aluminum may be better understood.

Analyses for total mercury also were conducted on the water samples. Two hypothesized sources of mercury in water are fallout from regional atmospheric concentrations of fossil fuel combustion products and/or mobilization of mercury from bedrock geological sources [2]. Mercury has been an important concern in the state since the discovery of elevated mercury residues in fish in several softwater lakes of the BWCA [23]. At that time no efforts were made to measure pH, alkalinity or mercury concentrations in the water. Studies conducted by Lander and Larsson in 1972 (as reported by Tomlinson et al. [24]) found that, in a series of interconnected lakes in Sweden, mercury concentrations in fish increased as the pH of the lakes decreased. More recently Scheider et al. [6] demonstrated that for walleye, fish of a given size have higher mercury content in lakes of low alkalinity ($\leqslant 15$ mg/L as $CaCO_3$) than those in higher-alkalinity waters. It is also important to note that at pH values <7.5, monomethyl mercury is the predominant methylated form. It is this form of mercury that is taken up by fish and subsequently bioaccumulated [24].

Mercury concentrations found in the water samples from our two surveys were very low compared to levels found in fish during the MDNR study of 1977 [23]. The concentrations in fish tissues ranged from 0.09 to 2.67 mg/L with a mean value of 0.84 mg/L (840 ppb). Our summer values ranged from <0.1 ppb (detection limit) to 0.39 ppb, with a mean of 0.15 ppb (Table 2). The fall values ranged from below the detection limit to 0.43 ppb (Table 3). The two lakes that exhibited these high concentrations had low alkalinities ($\leqslant 10$ mg/L and field pH values of 6.5). No discernible relationship was noted between pH and mercury concentrations in the water for either survey.

A comparison of the mercury data from the two surveys yields some interesting results. For the summer data, 24% of the lakes sampled had mercury concentrations below the detection limit, in contrast to 93% of the lakes in the fall survey. A comparison of the lakes sampled during both surveys revealed that for 16 of the 17 lakes, the fall concentrations were less than the summer concentrations (one lake remained the same). Further, only 1 of these lakes was below the detection limit in the summer, but 15 of the 17 were below the detection limit in the fall. No explanation is offered, at the present, for this observation.

COMPARISON WITH HISTORICAL DATA

Comparison of these current data wth historical data can give an indication of the long-term response of the lakes to acidic deposition. However, it is difficult to draw definite conclusions from such simple comparisons because of the inherent uncertainty involved with the data. Over the last 30 years analytical techniques have improved greatly, making comparisons with older data ambiguous. In addition, the seasonal, diurnal and locational variables are not controlled; but a sufficiently large sample, distributed over a wide area, can provide some insight into the effects of acid precipitation. Davis et al. [25] and Norton et al. [26] both examined historical surface water data and found increasing lake acidity that correlated with the history of acid precipitation.

One aspect of this study was to compare current pH and alkalinity values with data collected from past surveys. The MDNR has been performing limited chemical analysis on lake samples since 1936. We were able to find historical alkalinity data for 69% of the 1980 survey lakes in MDNR files. The mean age of these historical data was about 25 years (circa 1955). In some cases, information was available for more than one year; however, for this study we chose to select only the oldest reliable data to compare with the 1980 survey lakes. Unfortunately, historical pH values were only available for 20% of the cases. Since so few data were found, we excluded pH from consideration.

The results of our comparison are shown in Table 4. Based strictly on comparison of our current data (1980) with the oldest available data for 118 of the lakes, we found that the mean lake alkalinity decreased by 2.4 mg/L as $CaCO_3$. Approximately 75% of the lakes had lower alkalinity values in 1980 than for their first recorded samplings. The overall change in alkalinities in these lakes ranged from a countywide mean of +46.5 mg/L in southwestern Minnesota to -5.7 mg/L in extreme northeastern Minnesota. Individual lakes in central and northeastern Minnesota exhibited alkalinity decreases of up to 10-25 mg/L as $CaCO_3$. The MDNR alkalinity values were measured by fixed-endpoint titrations with brom-cresol green as an indicator. This proce-

Table 4. Historical Data Comparisons

County	Number of Lakes with Historical Data	Number of Lakes with Decrease in Alkalinity	Mean Decrease in Alkalinity[a]		Overall Change in Alkalinity[b]
			$\bar{x}$	Range	
Cook	34	31	6.2	0.4–13.6	–5.7
Lake	20	16	12.7	2.5–25.3	–4.6
St. Louis	33	21	5.4	0.5–14.0	+0.5
Carlton	5	4	6.6	1.0–13.8	–4.8
Pine	3	1	8.0	—	+0.5
Itasca	7	6	6.4	3.0–10.0	+12.9
Cass	5	4	6.6	4.7– 9.0	–4.0
Crow Wing	3	3	8.1	3.5–12.7	–8.0
Becker	3	2	9.5	6.0–13.0	+8.6
Washington	3	0	—	—	+33.3
Murry	2	0	—	—	+46.5
Totals	118	88	$\bar{x} = 7.4$		$\bar{x} = -2.4$

[a]Mean decrease in alkalinity for those lakes exhibiting an overall decrease (mg/L as $CaCO_3$).
[b]Mean change in alkalinity for all cases.

dure has a reported accuracy of ± 3.0 mg/L as $CaCO_3$. Although it is certain that the differences in analytical techniques may contribute to the observed trends, large decreases and increases in alkalinities are probably reflective of changes other than simply analytical procedures. For example, large increases in lake alkalinity with time may indicate increased erosion within the watershed, which is probably due to human activities such as agriculture and construction. Increases in lake alkalinity in the agricultural areas of western and southwestern Minnesota and in the resort-lakes area of central Minnesota are likely due to agriculture and construction related practices.

It is interesting to note that 43% of the lakes demonstrating decreasing alkalinities were clear and 38% were highly colored. Apparently, the trend toward decreasing alkalinity is not limited to either clear or bog-stained lakes. This is significant to note since bog lakes are naturally more acidic than clear lakes.

These findings are probably not highly significant when viewed alone. However, when viewed in conjunction with the historical patterns of acid precipitation, an interesting trend appears. Thornton et al. [1] found a dramatic transition from acidic to alkaline precipitation across Minnesota. The mean annual pH of rain and snow in extreme northeastern Minnesota was found to be 4.67, and the mean pH in extreme western Minnesota was 5.30. The uniform increase in precipitation acidity by a factor of 5 was attributed in part to the influences of calcareous prairie soil in the west coupled with the lack of exposed soils and the increased influence of anthropogenic emissions in the forested regions in northeastern Minnesota.

Figure 8 shows the results of our comparison study, geographically presented and compared to the current acidity of precipitation in Minnesota. The data were separated by county. The numbers in Figure 8 represent the percentage of the study lakes in each county that have demonstrated a decrease in alkalinity for each county. Superimposed over Figure 8 are isopleth lines which represent the general precipitation acidity in Minnesota [27].

Interestingly, the areas receiving the most acidic precipitation are the areas showing a higher number of lakes with decreasing alkalinities. Also, these areas, in general, show the greatest mean decrease in lake alkalinity with time. The two extreme northeastern counties, Lake and Cook, contained about half of our study lakes. Based on the results of our study, 80-90% of the lakes in these counties currently exhibit lower alkalinities than before, with an average decrease in alkalinity of 4-6 mg/L as $CaCO_3$. In contrast, few of the study lakes in western and southern Minnesota counties (farming areas which receive less acidic precipitation) show decreasing alkalinities and in fact have experienced an increase in alkalinity with time. The lakes we studied in central Minnesota showed a high percentage of declining alkalinities and a corresponding high mean decrease in lake alkalinity. Since these data

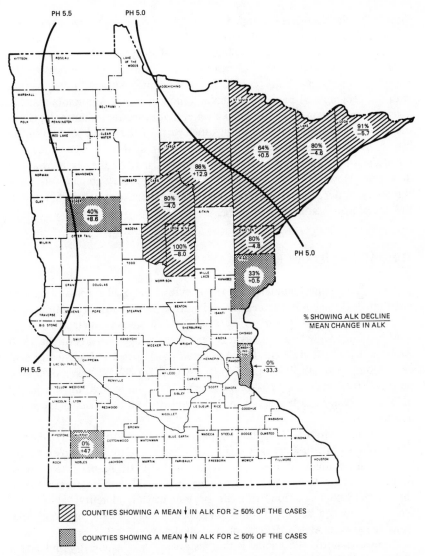

Figure 8. Historical alkalinity comparisons. pH isopleths represent annual precipitation pH [27].

represent only 6% of our comparison cases, the results for central Minnesota are probably not as significant as those for northeastern Minnesota.

Again, caution must be exercised before drawing any conclusions from this examination of trends in alkalinity data. The only accurate method of assessing trends is to duplicate completely the methods, sampling period and location of the historical researcher. Even then, such uncontrolled parameters as major changes in weather patterns and increased erosion in the watershed (from human activity) can affect temporal trend analysis.

However, the large number of cases showing decreased alkalinities and the magnitude of the alkalinity change in some counties viewed in conjunction with the geographical patterns of rain and snow acidity in Minnesota indicate a general decrease in alkalinity in lake water in northeastern Minnesota. Given the evidence of lake acidification in other parts of the world, and given the history of acid precipitation in North America, it is possible to conclude that the decreasing alkalinities in lakes of Minnesota may be the result of acidic depositon. However, it is impossible to prove this theory conclusively from this study, and it is also impossible to determine the degree of acidification or the rate of acidification in these areas.

SUMMARY AND CONCLUSIONS

Lake monitoring conducted in 1980 reveals that acid sensitive lakes are distributed throughout the central and northeastern portion of Minnesota. This area corresponds to the region of the state that was predicted to be acid-sensitive from knowledge of bedrock and soil geology. Results of the 1980 study indicate that 42% of the lakes could be classified as sensitive to acid deposition (total alkalinity 0-15 mg/L as $CaCO_3$) and 25% as potentially sensitive (total alkalinity 16-30 mg/L as $CaCO_3$). In general, the sensitivity of lakes in Minnesota is greatest in the northeast, and decreases to the south and west.

About one-third of the lakes we sampled were found to be highly colored. The sensitivity of these lakes is difficult to assess, due to the complex nature and organic content of bog waters. Potentially toxic Al–pH values were found in a few colored lakes, but toxicity is probably not a concern due to the possible complexation with the organic matter. Mercury values were mostly below the detection limits. In general, there was no apparent relationship between Al and Hg with pH and alkalinity.

Comparisons of 1980 data with "historical" lake chemistry indicate a slight overall decrease in alkalinity in the last 10–45 years. Although changes in analytical techniques may account in part for this difference, we found significant decreases in alkalinity (10–25 mg/L as $CaCO_3$) for many individual lakes in central and northeastern Minnesota. The highest occurrences of alkalinity decline correlate well with the areas receiving the most acidic precipitation. Although we found no evidence of chronic lake acidification in Minnesota, these data suggest that the acidification process may be occurring.

Future monitoring efforts should address more closely the interrelationship of watershed geology and water chemistry to allow for better prediction of acid sensitive aquatic resources. In addition, future studies should consider employing older analytical techniques as well as modern methods to allow a more quantitative comparison of current and historical data.

REFERENCES

1. Thornton, J. D., et al. "Trace Metal and Strong Acid Composition of Rain and Snow in Northern Minnesota," in *Atmospheric Input of Pollutants to Natural Waters*, S. J. Eisenreich, Ed. (Ann Arbor, MI: Ann Arbor Science Publishers, Inc., 1981).
2. Glass, G. E., and O. L. Loucks, Eds. "Impacts of Airborne Pollutants on Wilderness Areas Along the Minnesota-Ontario Border," U.S. EPA Report 600/3-80-044 (1980).
3. "An Inventory of Minnesota Lakes," Minnesota Department of Natural Resources Bulletin No. 25 (1968).
4. Moyle, J. B. "Some Chemical Factors Influencing the Distribution of Aquatic Plants in Minnesota," *Am. Midl. Nat.* 34:402-420 (1945).
5. "Methods for Chemical Analysis of Water and Wastes," U.S. EPA/600-4-79-020 (1979).
6. Scheider, W. A., D. S. Jeffries and P. J. Dillon. "Effects of Acidic Precipitation on Precambrian Freshwaters in Southern Ontario," *J. Great Lakes Res.* 5:(1)45-51 (1979).
7. Henriksen, A. "A Simple Approach for Identifying and Measuring Acidification of Freshwater," *Nature* 278:(15)542-545 (1979).
8. Stumm, W., and J. J. Morgan. *Aquatic Chemistry* (New York: John Wiley and Sons, Inc., 1970), pp. 118-160.
9. Kramer, J. R. "Precise Determination of Low Alkalinities Using the Modified Gran Analysis, an Inexpensive Field Procedure," Department of Geology, McMaster University, Hamilton, Ontario (1980).
10. *Standard Methods for the Examination of Water and Wastewater,* 14th Ed. (New York: American Public Health Association, 1976).
11. Glass, G. E., N. R. Glass and P. J. Rennie. "Effects of Acid Precipitation," *Environ. Sci. Technol.* 13:(11)1350-1355 (1979).
12. Beamish, R. J. "Acidification of Lakes in Canada by Acid Precipitation and the Resulting Effects on Fishes," *Water Air Soil Poll.* 6:501-504 (1976).
13. Wetzel, R. G. *Limnology* (Philadelphia, PA: W. B. Saunders, 1975), 743 pp.
14. Dickson, W. "The Acidification of Swedish Lakes," Institute of Freshwater Research, Drottningholm, Sweden, Report No. 54, 8-20 (1975).
15. Johannessen, M. "Aluminum, a Buffer in Acidic Waters?" in *Ecological Impact of Acid Precipitation*, D. Drablos and A. Tollan, Eds., Proceedings of an international conference, Sandefjord, Norway, March 11-14, SNFS Project (1980), pp. 222-223.

16. Dickson, W. "Properties of Acidified Waters," in *Ecological Impact of Acid Precipitation*, D. Drablos and A. Tollan, Eds., Proceedings of an international conference, Sandefjord, Norway, March 11–14, SNFS Project (1980), pp. 75–83.

17. Wright, R. F., and E. T. Gjessing. "Changes in the Chemical Composition of Lakes," *Ambio* 5:(5–6)219–223 (1976).

18. Cronan, C. S., and C. L. Schofield. "Aluminum Leaching Response to Acid Precipitation: Effects on High-Elevation Watersheds in the Northeastern U.S.," *Science* 204:(20)304–306, (1979).

19. Wright, R. F., et al. "Acidified Lake Districts of the World: A Comparison of Water Chemistry of Lakes in Southern Norway, Southern Sweden, Southwestern Scotland, the Adirondack Mountains of New York, and Southern Ontario," in *Ecological Impact of Acid Precipitation*, D. Drablos and A. Tollan, Eds., Proceedings of an international conference, Sandefjord, Norway, March 11–14, SNFS Project (1980), pp. 377–379.

20. Muniz, I. P., and H. Leivestad. "Toxic Effects of Aluminum on the Brown Trout, *Salmo trutta*, L.," in *Ecological Impact of Acid Precipitation*, D. Drablos and A. Tollan, Eds., Proceedings of an international conference, Sandefjord, Norway, March 11–14, SNFS Project (1980), pp. 320–321.

21. Driscoll, C. T. "Aqueous Speciation of Aluminum in the Adirondack Region of New York State, U.S.A.," in *Ecological Impact of Acid Precipitation*, D. Drablos and A. Tollan, Eds., Proceedings of an international conference, Sandefjord, Norway, March 11–14, SNFS Project (1980), pp. 214–215.

22. Muniz, I. P., and H. Leivestad. "Acidification—Effects on Freshwater Fish," in *Ecological Impact of Acid Precipitation*, D. Drablos and A. Tollan, Eds., Proceedings of an international conference, Sandefjord, Norway, March 11–14, SNFS Project (1980), pp. 84–92.

23. "Mercury Levels in Fish from Eleven Northeastern Minnesota Lakes, 1977," Minnesota Department of Natural Resources. Investigational Report No. 348 (1978).

24. Tomlinson, G. H., et al. "The Role of Clouds in Atmospheric Transport of Mercury and Other Pollutants: 1) The Link Between Acid Precipitation, Poorly Buffered Waters, Mercury, and Fish," in *Ecological Impact of Acid Precipitation*, D. Drablos and A. Tollan, Eds., Proceedings of an international conference, Sandefjord, Norway, March 11–14, SNFS Project (1980), pp. 134–136.

25. Davis, R. B., et al. "Acidification of Maine Lakes," *Verh. Internat. Verein. Limnol.* 20:532–537 (1978).

26. Norton, S. A., et al. "Changing pH and Metal Levels in Streams and Lakes in the Eastern United States Caused by Acidic Precipitation," U.S. EPA Lakes Restoration Conference, Portland, ME, 1980.

27. Gibson, J. H. and Baker, C. V. "NADP First Data Report July 1978 through February 1979," National Resurce Ecology Laboratory, Colorado State University, Fort Collins, CO (1979).

OTHER SOURCES

1. Jeffries, D. S., C. M. Cox and P. J. Dillon. "Depression of pH in Lakes and Streams in Central Ontario During Snowmelt," *J. Fish Res. Bd. Can.* 36:640–646 (1979).
2. Minnesota Department of Natural Resources. Unpublished data.
3. Schofield, C. L. "Effects of Acid Rain on Lakes," in *Acid Rain– Proceedings of ASCE National Convention,* C. G. Gunnerson and B. E. Willard, Eds. (1979), pp. 55–69.
4. *Standard Methods for the Examination of Water and Wastewater,* 12th ed. (New York: American Public Health Association, 1965).

CHAPTER 10

CHRONOLOGY OF ATMOSPHERIC DEPOSITION OF ACIDS AND METALS IN NEW ENGLAND, BASED ON THE RECORD IN LAKE SEDIMENTS

S. E. Johnston and S. A. Norton

Department of Geological Sciences
University of Maine at Orono
Orono, Maine 04469

C. T. Hess

Department of Physics
University of Maine at Orono
Orono, Maine 04469

R. B. Davis and R. S. Anderson

Department of Botany and Plant Pathology
and the Institute for Quaternary Studies
University of Maine at Orono
Orono, Maine 04469

INTRODUCTION

The establishment of the chronology of sediments deposited during the last few hundred years has enabled researchers to understand a great deal about rates of natural limnological and oceanographic processes. Under the right circumstances this dating may be linked to terrestrial and atmospheric processes. In profundal lake sediments, dating has been accomplished by counting varves (annual pairs of laminations) [1], stratigraphic markers of known age (e.g., catastrophic erosion and subsequent accelerated deposition, pulse inputs of volcanic ash, pollen [2], and chemically immobile radionuclides or other pollutants [3], and radioactive decay of unsupported

177

nuclides (e.g., Pb^{210}) [3]. Mechanical mixing of sediments, resuspension of sediments, "wash-in" from littoral areas, delayed wash-in from the drainage basin, bioturbation, sediment diagenesis and non-steady-state conditions may affect these dating methods.

For the past five years we have been evaluating the influence of anthropogenic activities on watersheds and lakes, as recorded by lake sediments [2]. Recently we have been concerned about the chronology of perturbations of the terrestrial and aquatic ecosystems caused by atmospheric deposition of acids and metals. To eliminate other effects such as erosion plus accelerated deposition, we have studied relatively pristine lake ecosystems in remote regions. Several dating techniques (e.g., varves) cannot be employed in our studies because of bioturbated sediments. Consequently we designed a study to evaluate the Cs^{137} and Pb^{210} dating techniques against other methods. The most appropriate method could then be utilized in our paleolimnological reconstruction of the chronology of polluted precipitation and its effects.

METHODS

Sediment sections for the comparative chronology studies were obtained from the deepest part of Conroy Lake and Rideout Pond (Table 1) using a "frozen finger" sampler utilizing the design of Swain [4]. Ledge Pond sediments were retrieved with a 10-cm diam stationary piston corer [5].

The "frozen finger" cores were kept frozen and sectioned in the laboratory with a saw after "varve" counts were performed. The Ledge Pond core was sectioned in the field. After sectioning, all sediments were stored in plastic bags in the dark at $4°C$. Further processing was as follows:

1. Mechanical homogenization was accomplished by kneading the sediment in the bag.
2. Volumetric aliquots were removed for analyses of pollen, diatoms and *Cladocera* (not reported herein). Pollen slides were prepared and analyzed according to methods of Faegri and Iverson [6].
3. The remainder of the sediment was handled as follows:
    a. It was oven dried at $110°C$ for 24 hr (water content).
    b. A separate split of 3a was combusted at $550°C$ in a muffle furnace for 3 hr to determine "oxidizable hydrocarbons."
    c. It was crushed and packed into plastic Petri dishes. Cesium-137 activities were measured using a Tracor Northern Econ II gamma ray spectrometer and a liquid N_2-cooled lithium-drifted germanium detector. Count time ranged from 20,000 to 40,000 sec.
    d. The same sample (3c) was used for volatizing and collecting ^{210}Pb following a technique slightly modified from that of Eakins and Morrison [7] (further details are given in Reference 8). Controlled studies indicate that this procedure does not volatize the elements of interests (see 3e). ^{208}Pb and ^{210}Pb were determined with a Tracor Northern 1710 alpha

Table 1. Characteristics of Study Lakes in Maine

	Water Depth at Coring Site (m)	Area of Lake (ha)	Area of Drainage (ha)	Altitude (m)	Midsummer pH	Disturbances
Conroy (Meromictic)	33	9.1	740	148	7.1	Deforestation and agriculture
Rideout (Meromictic)	21	0.89	316	320	7.2	Deforestation
Ledge (Temporarily stratified)	7.3	~4	~50	893	4.5	Past: selective cutting. Now: undisturbed.

ray spectrometer with an Ortec surface barrier alpha detector. Count times were 5000 to 10,000 sec.

e. The same sample (3d) was prepared for bulk chemical analysis using the methods of Buckley and Cranston [9] except that the digestion was done directly in linear polyethylene bottles. Analyses for major elements were made using a Perkin-Elmer #703 atomic absorption spectrophotometer (flame); minor elements were determined with an H.G.A. 2200 graphite furnace coupled with the 703.

RESULTS

Stratigraphic markers of postulated age were present in Conroy Lake and Rideout Pond sediments (both lakes have profundal sediments with no bioturbation). Two distinct allochthonous bands of clay-sized material were present at 5.5 and 10.5 cm in the Conroy Lake sediment section. The upper band is thought to have been caused by high atmospheric precipitation events in 1977, which resulted in exceptional erosion of soil and deposition of sediment. The rediversion and subsequent scouring of an inlet in 1966 is thought to have led to the formation of the stratigraphic marker at 10.5 cm. A distinct stratigraphic marker in the Rideout Pond sediment (6.5 cm) is postulated to be a result of lumbering in 1950.

At Conroy Lake, sediment dates based on the interpretation of lamination pairs (couplets) as annual (varves) were too old when compared to the position of the 1977 and 1966 stratigraphic markers (Figure 1). This suggests that more than one couplet was deposited during one year. At Rideout Pond, a down-core extrapolation of sediment dates determined from couplets (interpreted as varves) in the upper 4.5 cm are in agreement with the 1950 stratigraphic marker (Figure 2).

Sediment dates determined from the 1954 sharp rise in ^{137}Cs activity were in close agreement with dates derived from the stratigraphic markers in Conroy Lake and Rideout Pond, whereas the sediment dates using the 1963 peak in ^{137}Cs activity, by comparison, were too old. The lag, which is about five years in the deposition of the ^{137}Cs (1963) is postulated to be the result of a delayed terrestrial and littoral contribution of ^{137}Cs to the profundal sediments of Conroy Lake and Rideout Pond, resulting in a ^{137}Cs peak closer to the sediment surface than would be normally expected from direct atmospheric input.

The ^{210}Pb dating models were in close agreement with dates determined from the 1977 and 1966 stratigraphic markers in Conroy Lake. In Rideout Pond, sediment dates using the constant input concentration (CIC) model [3], and the 1950 stratigraphic marker were also in good agreement, but the constant initial flux (CIF) model [10] gives dates that are too old (Figure 2).

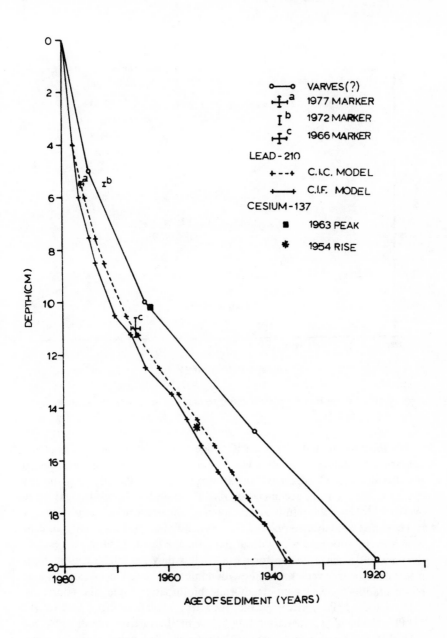

Figure 1. Comparative sediment dating techniques for Conroy Lake, Maine.

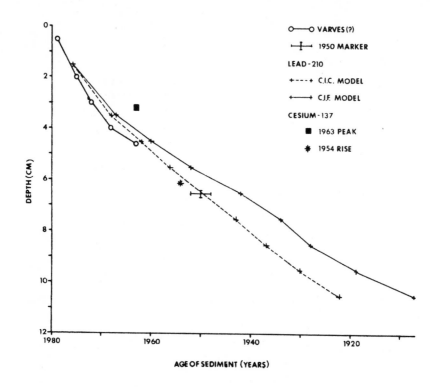

Figure 2. Comparative sediment dating techniques for Rideout Pond, Maine.

We conclude that the ^{210}Pb CIC model and 1954 rise in ^{137}Cs activity are accurate dating methods when compared to stratigraphic markers of postulated age in Conroy Lake and Rideout Pond. The ability to monitor a change in the sediment accumulation rate using the ^{210}Pb profile is preferred over the ^{137}Cs method. In the latter method, the sediment accumulation rate is dependent on only two points (1954 rise and 1963 peak). Lamination pairs may form by processes which occur more than or less than once per year and thus are not always reliable for dating unless they are established by other methods to be true varves. In lakes with nonlaminated sediments, steady-state sedimentation coupled with steady-state bioturbation affect the uppermost part of the ^{210}Pb activity profile but have only a minimal effect on the ^{210}Pb calculated sedimentation rate below the bioturbation level. However, bioturbation would smear (upward and downward) pulse inputs or stratigraphic markers; where tubificids dominate the herpobenthos, the remaining peaks (e.g., 1963 ^{137}Cs) would be buried (moved downward) by older sediment [11]. Assignment of absolute age to these markers will yield sediment accumulation rates which are too high. Bench-mark horizons obviously give correct age for the sediment itself but not for the sedimentation rate. Bio-turbation by tubificids and delayed wash-in may have compensating effects on the use of stratigraphic markers for dating. With increasing age of strati-graphic markers, they converge on the correct age/sedimentation rate curve derived from ^{210}Pb.

As an example of the application of these dating methods to nonlaminated sediment, we give results from Ledge Pond, a pristine, acidic lake in Western Maine (Table 1). These results are typical for such lakes in northern New England. The undisturbed watershed is vegetated largely by spruce-fir forest. Figure 3 shows the ^{137}Cs and ^{210}Pb profiles. The ^{137}Cs profile does not resemble the pattern mimicking the profile expected from atmospheric deposition records [12]. There is no sharp rise in concentration (ca. 1954), no pronounced peak in activity (ca. 1963), and no decrease to near-zero values for the 1970s. This may be caused by bioturbation and delayed and prolonged wash-in of ^{137}Cs from the terrestrial ecosystem as well as from shallower areas of the lake which confuse the simple atmospheric deposition record. Additionally, it appears likely that ^{137}Cs is considerably more chem-ically mobile on the terrestrial scene [13] as well as in some lake sediments than has previously been thought (see, e.g., Reference 14).

Figure 4 compares sediment age as determined by ^{210}Pb for Ledge Pond. Pollen ages (not illustrated) are based on the history of settlement (at lower elevation) and lumbering, but are somewhat imprecise because of long-range transport of pollen. If anything, the assigned ages (7 cm = 1870, hemlock and pine pollen decreases associated with lumbering; 11 cm = 1810, weed pollen increases associated with settlement) are 25 to 30 years older than those predicted by extrapolation of the ^{210}Pb curves. The ^{137}Cs age departs markedly from ^{210}Pb ages. If the 1954 "rise" is taken to be at 3.75 cm, the ^{137}Cs age is 6-8 years ($\sim 25\%$) younger than the ^{210}Pb age. If the 1954 "rise" is taken to be at 6.0 cm (Figure 3), then the ^{137}Cs age is 39-41 years younger than the ^{210}Pb age. The 1963 ^{137}Cs peak is too indistinct to use. It is noteworthy that the 1954 ^{137}Cs "rise" age for sediment agrees with the ^{210}Pb ages in the two lakes with laminated sediments (minimal bio-turbation) but not in the bioturbated sediments of Ledge Pond. If the ^{210}Pb age curve (both CIC and CIF methods) is extrapolated downward with a compaction correction, the dated onset of cultural pollen agrees with the ^{210}Pb age reasonably well. Thus we have based our sediment chronology on ^{210}Pb and pollen.

Total sediment Pb and Zn concentrations are shown on Figure 4. They increase concurrently at 9 cm (about 1875). This date approximately agrees with other estimates for the age of increasing atmospheric deposition of metals in other lakes in New England [15] and elsewhere in the eastern United

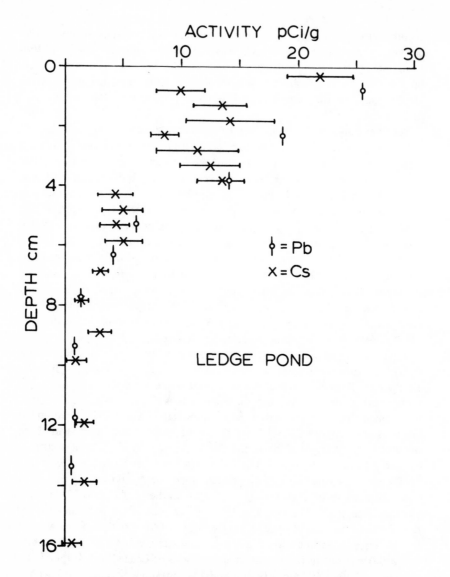

Figure 3. [210]Pb and ^{137}Cs profiles for sediment from Ledge Pond, Maine.

States. Sediments in historically acidified lakes (pH ⩽ 5.5) consistently have decreasing concentrations of Zn toward the top of the sediment [15]. This is believed to be due partly to accelerated leaching of Zn from soils (precursors of sediment [13, 16]) by increasingly acidic precipitation. In Ledge

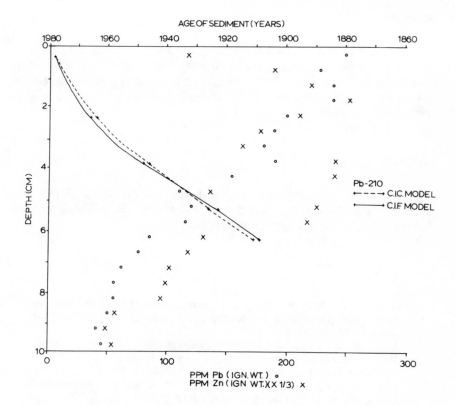

Figure 4. Sediment age (^{137}Cs, ^{210}Pb and pollen) and chemistry for sediment from Ledge Pond, Maine.

Pond, sediment peak Zn concentrations are reached at about 3.5 cm (~ 1950), indicating that accelerated leaching of Zn from soils had commenced by that time. Lakes similar to Ledge Pond show decreases in concentrations of Fe, Mn, Mg, Ca and Zn in sediment [16], indicating that accelerated leaching of these elements from soils had commenced around some New England lakes as early as 1930-1940. (By implication, the pH of surface waters should have been decreased.) Pb and Zn concentrations both increase toward the top of the sediment in Rideout and Conroy, both of which have higher pH and more alkaline soils in their drainage basins.

SUMMARY

Although historic precipitation chemistry data are sparse, there is a general relationship between decreasing pH and increasing atmospheric deposition of metals (Pb and Zn, at least) [17]. Thus, based on the data presented herein, acidified precipitation and elevated metal deposition were present as early as 1875 in New England.

ACKNOWLEDGMENTS

This research was supported by The Society of the Sigma Xi (S. E. Johnston) and the U.S. National Science Foundation (R. B. Davis, S. A. Norton and C. T. Hess) Grant #DEB-7810641.

REFERENCES

1. Renberg, I., "Paleolimnology and Varve Counts of the Annually Laminated Sediments of Lake Rudetjarn, Northern Sweden," in *Early Norrland II*, (1978), pp. 64–91.
2. Davis, R. B., and S. A. Norton. "Paleolimnologic Studies of Human Impact on Lakes in the United States, with Emphasis on Recent Research in New England," *Pol. Arch. Hydrobiol.* 25:99–115 (1978).
3. Robbins, J. A., and D. N. Edgington. "Determination of Recent Sedimentation Rates in Lake Michigan Using Pb–210 and Cs–137," *Geochim. Cosmochim. Acta* 39:285–304 (1975).
4. Swain, A. M. "A History of Fire and Vegetation in Northeastern Minnesota as Recorded in Lake Sediments," *Quat. Res.* 3:383–396 (1973).
5. Davis, R. B., and R. W. Doyle. "A Piston Corer for Upper Sediment in Lakes," *Limnol. Oceanog.* 14:643–648 (1969).
6. Faegri, K., and J. Iverson. *Textbook of Pollen Analysis* (New York: Hafner Publishing Company, 1975), 237 pp.
7. Eakins, J. D., and R. T. Morrison. "A New Procedure for the Determination of Lead–210 in Lake and Marine Sediments," *Intern. Appl. Rad. Iso.* 29:531–536 (1978).
8. Johnston, S. E. "A Comparison of Dating Methods in Laminated Lake Sediments in Maine," MS Thesis, University of Maine at Orono (1981), 79 pp.
9. Buckley, D. E., and R. E. Cranston. "Atomic Absorption Analysis of 18 Elements from a Single Decomposition of Aluminosilicate," *Chem. Geol.* 7:273–284 (1971).
10. Appleby, P. G., and F. Oldfield. "The Calculation of Lead–210 Dates Assuming a Constant Rate of Supply of Unsupported Pb–210 to the Sediment," *Catena*, 5:1–8 (1978).

11. Davis, R. B. "Tubificids Alter Profiles of Redox Potential and pH in Profundal Lake Sediments," *Limnol. Oceanog.* 19:342–346 (1974).

12. Toonkel, L. E. Environmental Measurements Laboratory, Environmental Quarterly: U.S. Department of Energy (1980).

13. Hanson, D. W. "Acidic Precipitation-Induced Chemical Changes in Subalpine Fir Forest Organic Soil Layers," MS Thesis, University of Maine at Orono (1980), 90 pp.

14. Francis, C. W., and F. S. Brinkley. "Preferential Adsorption of Cs-137 to Micaceous Minerals in Contaminated Freshwater Sediment," *Nature*, 260:511–513 (1976).

15. Norton, S. A., D. W. Hanson and R. J. Campana. "The Impact of Acidic Precipitation and Heavy Metals on Soils in Relation to Forest Ecosystems," in *Effects of Air Pollutants on Mediterranean and Temperate Forest Ecosystems*, Miller, R. R., Ed., U.S. Department of Agriculture Gen. Tech. Rept. PSW–43 (1980), pp. 152–157.

16. Norton, S. A., "Changing pH and Metal Levels in Streams and Lakes in the Eastern United States Caused by Acidic Precipitation," Proceedings of the EPA Conference on Lake Restoration (in press).

17. Galloway, J. N., et al. "Toxic Substances in Atmospheric Deposition: A Review and Assessment," U.S. EPA Rept. 560/5-80-001 (1981), pp. 19–82.

CHAPTER 11

RELATIONSHIPS OF CHEMICAL WET DEPOSITION TO PRECIPITATION AMOUNT AND METEOROLOGICAL CONDITIONS

Gilbert S. Raynor and Janet V. Hayes

Atmospheric Sciences Division
Brookhaven National Laboratory
Upton, New York 11973

INTRODUCTION

Hourly precipitation samples that have been collected at Brookhaven National Laboratory (BNL) during all precipitation events since June 1976 were analyzed for concentrations of major ions. Meteorological data were recorded for each sample period. Previous papers [1-4] described the relationships between precipitation chemistry and meteorological conditions and showed that concentrations of all measured ions in BNL precipitation vary systematically with season, synoptic situation, precipitation type, precipitation rate and other parameters. A recent paper [5] analyzed wet deposition in a similar way. This chapter relates wet deposition amounts for selected meteorological conditions to precipitation amounts under the same conditions. Wet deposition amounts elsewhere in eastern and central North America have been reported [6-13], but none have related wet deposition amounts to the variables considered in this study.

METHODS

More than 2800 precipitation samples collected with an automatic sequential sampler designed and constructed at BNL [14] from June 1976 through May 1980 were used in this study. They were analyzed for pH, conductivity and concentrations of sulfate, nitrogen in nitrate plus nitrite, nitrogen in ammonium, sodium and chloride. Hydrogen ion concentrations were computed from the pH measurements. Deposition of each chemical per unit area was calculated for each sample. Average deposition was computed for subsets of the data classified by time and meteorological parameters including year, season, month, synoptic type, precipitation type, precipitation rate, wind direction, wind speed and temperature. Results are presented as deviations from the percentages that would be expected if chemical wet deposition were proportional to precipitation amount. Deviations were computed for each species as the percentage of the total wet deposition in each subclass minus the percentage of the total precipitation in that subclass. Thus, a positive deviation indicates a greater than expected wet deposition of that chemical in that particular subclass. Proportionality between wet deposition amount and precipitation amount can be considered a first approximation only. Likens et al. [10], for instance, reported that the average input of SO_4^{2-} and NO_3^- to the Hubbard Brook watershed was proportional to the amount of precipitation on an annual basis, but no data have been available to describe the relationship under selected conditions.

RESULTS

Our data show that chemical wet deposition of all ions measured is a function of concentration in the precipitation as well as the precipitation amount. This was expected because of the variability in concentration with time and meteorological conditions found previously. However, the data presented here document the relative importance of concentration and precipitation amount in determining wet deposition amounts. They also identify the conditions under which chemical wet deposition is most disproportionate to precipitation amount. This information should have general application in studies of the effects of acidic precipitation on plants, soils and both ground and surface waters.

Significant differences were found between the four years included in this study (Figure 1). No two species showed the same year-to-year patterns in deviation of wet deposition amounts from precipitation amounts. Hydrogen ion had positive deviations in the first two years and negative deviations in the second two, but sulfate ion had a positive deviation only in the first year.

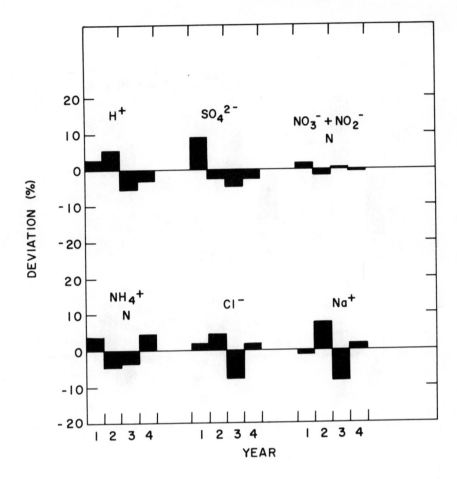

Figure 1. Deviations of chemical wet deposition percentages from precipitation percentages by year.

Nitrogen in nitrate plus nitrite was near the expected value each year but nitrogen in ammonium showed larger excursions. Except for the first year, chloride and sodium deviations were in phase. These patterns result from average differences in concentration in the precipitation. Sulfate concentrations, for instance, were high the first year.

When classified by season (Figure 2) deviation of hydrogen and sulfate ion deposition from expected amounts is most positive in summer when precipitation amounts are lowest and most negative in the winter when precipitation is heaviest. This results from higher concentrations in summer

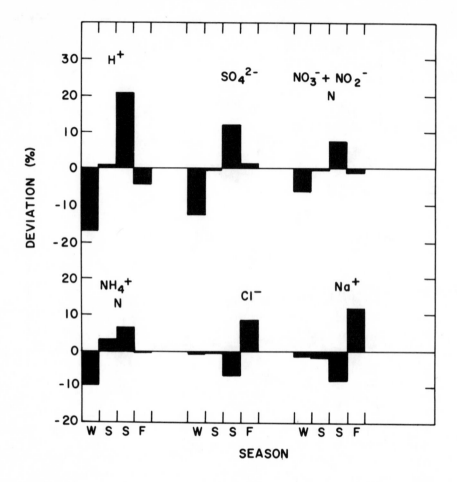

Figure 2. Deviations of chemical wet deposition percentages from precipitation percentages by season.

precipitation and lower concentrations in the winter. Nitrogen in both nitrate plus nitrite and in ammonium shows a similar, but less extreme, pattern, indicating less variability in concentration with season. Thus, the four species of presumed largely anthropogenic or terrestrial origin have similar seasonal patterns apparently related to the more highly polluted, slow moving air masses of the warmer season. In contrast, deviations of sodium and chloride ion concentrations are most negative in summer and most positive in fall. These species are derived mainly from sea salt and, as will be shown later, their patterns are caused primarily by seasonal differences in wind speed.

Distributions of deviations by month (Figure 3) reinforce the seasonal findings. The first four species show largely positive deviations during the warmer months and negative deviations in the winter. Chloride and sodium have opposite patterns.

Deviations of the four anthropogenic pollutants are most positive during cold front and squall line precipitation when classified by synoptic type (Figure 4) and most negative during warm front cases. Chloride and sodium have essentially opposite patterns. Most cold front and squall line precipitation falls in the relatively polluted air in the warm sector ahead of a cold front while warm front precipitation falls through colder and cleaner air masses. In addition, trajectories to BNL are often at least partially over the ocean during warm front precipitation.

Figure 5 shows that deviation from expected amounts of hydrogen and sulfate ion deposition is most positive during thundershowers and rain showers and strongly negative during rain when classified by precipitation type. These results are related to those above since showers and thundershowers usually occur with cold front and squall line passages, and rain is strongly associated with warm fronts. Note that the deviation of nitrogen is more positive with rain showers than with thundershowers, in contrast to the two former species, and has a secondary positive peak with snow. This conforms to previous findings of high nitrogen concentration in snow and at cold temperatures. Again, sodium and chloride have almost inverse patterns.

When classified by precipitation rate (Figure 6) hydrogen ion concentration shows a positive deviation only at the heaviest precipitation rate, which most often occurs during thundershowers. The sulfate deviation, however, is most positive at the lowest rate and most negative at intermediate rates, in close agreement with the pattern for nitrogen. The significance of these differences needs further study, but they may be partially caused by greater dilution of available airborne materials at higher precipitation rates. This is also probably true of sodium and chloride, which show patterns somewhat similar to those of nitrogen.

Classification of the data by wind direction (Figure 7) shows that the four species of anthropogenic origin have positive deviations primarily with winds from westerly directions. The indicated higher concentrations may originate in the New York City metropolitan area or in more distant industrial regions. Later studies are expected to resolve this uncertainty. The two marine species have positive deviations only with onshore winds (E-S), further evidence of their origin.

Large deviations in both directions are found when the cases are sorted by wind speed (Figure 8). The reasons for the positive deviations in the first four species in the 3- to 5-m/sec class are not known except that this is a common summer wind speed. The large positive deviations in sodium and

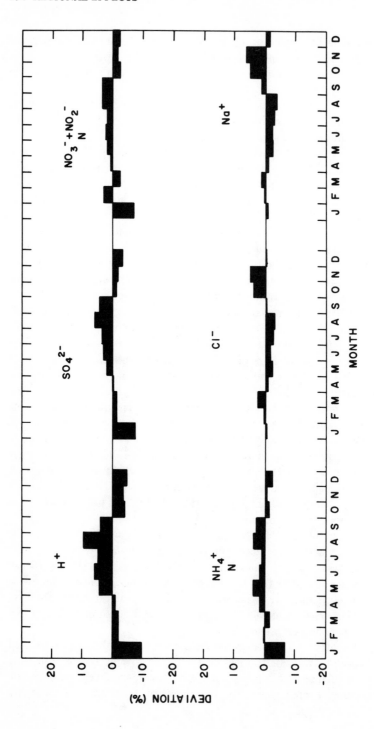

Figure 3. Deviations of chemical wet deposition percentages from precipitation percentages by month.

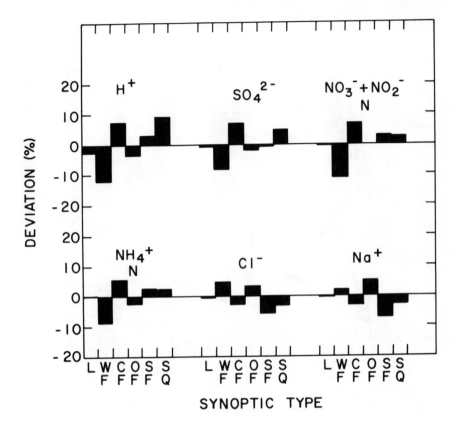

Figure 4. Deviations of chemical wet deposition percentages from precipitation percentages by synoptic type as follows: L, low; WF, warm front; CF, cold front; OF, occluded front; SF, stationary front; SQ, squall line.

chloride at high wind speeds are obviously a result of sea salt particle production by wave action, bubble bursting and breaking of the surf along the shore. The large negative deviations of the other species at high wind speeds may result from greater diffusion of available pollutants at these speeds.

The deviation distributions by temperature class (Figure 9) are similar to those by season (Figure 2). Deviations of the anthropogenic species are generally positive at warm temperatures and negative with moderate temperatures. It is obvious that temperature is not a major determinant of relative sodium and chloride wet deposition since all deviations are relatively small.

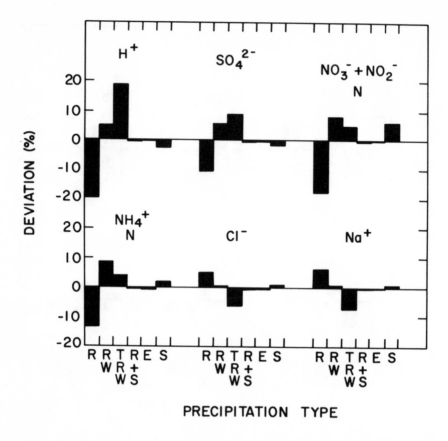

Figure 5. Deviations of chemical wet deposition percentages from precipitation percentages by precipitation type as follows: R, rain; RW, rain shower; TRW, thundershower; R+S, rain and snow; E, sleet; S, snow.

SUMMARY

Amounts of chemical wet deposition are greater than expected when related to the amount of precipitation under the following conditions:

- *hydrogen ion:* summer season, with cold fronts and squall lines, with thundershowers and rain showers, at high precipitation rates, with westerly wind directions, at wind speeds of 3–5 m/sec and at high temperatures;
- *sulfate ion:* similar except with low instead of high precipitation rates;
- *nitrogen in nitrate plus nitrite:* similar to sulfate except for a secondary peak with snow and cold temperatures;

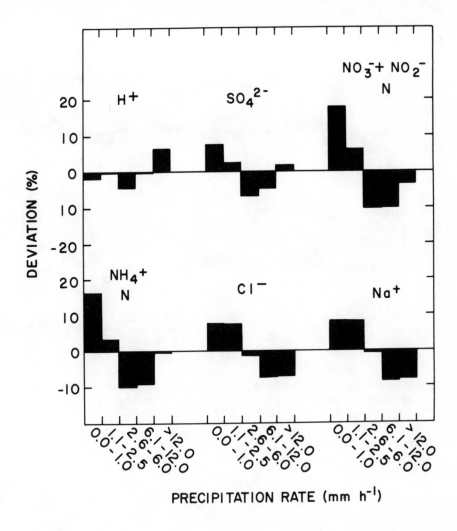

Figure 6. Deviations of chemical wet deposition percentages from precipitation percentages by precipitation rate.

- *nitrogen in ammonium:* similar to sulfate;
- *chloride and sodium:* fall season, with warm and occluded fronts, with rain at low precipitation rates, with onshore wind directions, with high wind speeds and at moderate temperatures.

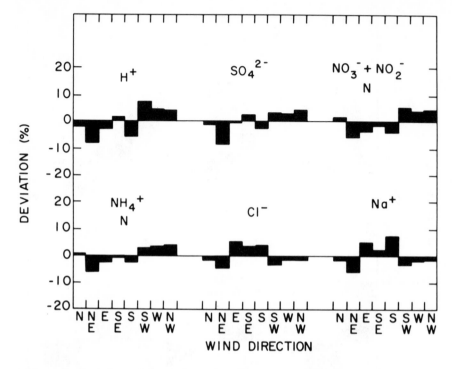

Figure 7. Deviations of chemical wet deposition percentages from precipitation percentages by wind direction.

Amounts are less than expected with the following conditions:

- *hydrogen ion:* winter season, with warm fronts, with rain, with moderate precipitation rates, with northeasterly and southerly winds, with high wind speeds and with cold temperatures;
- *sulfate ion:* similar to hydrogen except at precipitation rates of 2.6–12.0 mm/hr;
- *nitrogen in nitrate plus nitrite:* similar to sulfate except for temperature;
- *nitrogen in ammonium:* similar to sulfate;
- *chloride and sodium:* summer season, with stationary fronts, with thundershowers, with high precipitation rates, with northerly and westerly wind directions, with wind speeds of 3–5 m/sec and with temperatures of 0–5 and 15–20°C.

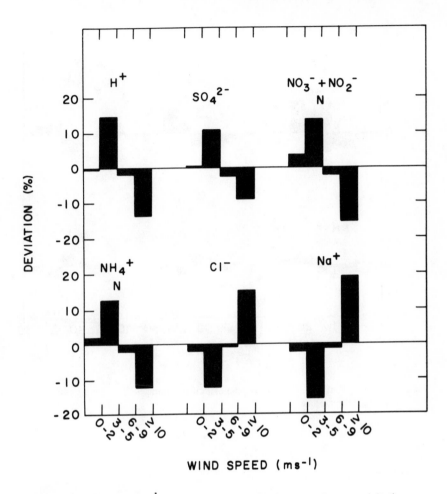

Figure 8. Deviations of chemical wet deposition percentages from precipitation percentages by wind speed.

ACKNOWLEDGMENTS

This research was performed under the auspices of the U.S. Department of Energy under Contract No. DE-AC02-76CH00016 and the U.S. Environmental Protection Agency Contract No. 79-DX-0533.

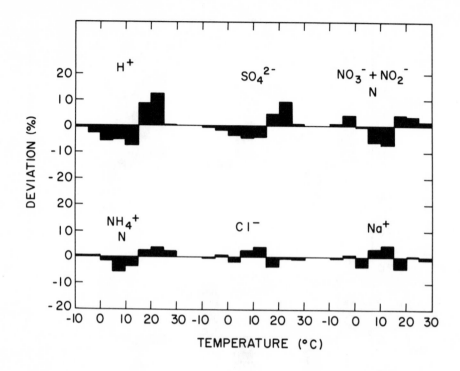

Figure 9. Deviations of chemical wet deposition percentages from precipitation percentages by temperature.

REFERENCES

1. Raynor, G. S., and J. V. Hayes. "Analytical Summary of Experimental Data from Two Years of Hourly Sequential Precipitation Samples at Brookhaven National Laboratory," Report BNL 51058, Brookhaven National Laboratory, Upton, NY (1979).
2. Raynor, G. S., and J. V. Hayes. "Sulfate, Nitrate plus Nitrite and Ammonium Ion Concentrations in Central Long Island, New York Precipitation in Relation to Meteorological Variables," Report BNL 28261, Brookhaven National Laboratory, Upton, NY (1980).
3. Raynor, G. S., and J. V. Hayes. "Chloride and Sodium Ion Concentrations in Central Long Island, New York Precipitation in Relation to Meteorological Variables," Report BNL 28333, Brookhaven National Laboratory, Upton, NY (1980).
4. Raynor, G. S., and J. V. Hayes. "Acidity and Conductivity of Precipitation on Central Long Island, New York in Relation to Meteorological Variables," *Water Air Soil Poll.* 15: 229–245 (1981).

5. Raynor, G. S., and J. V. Hayes. "Variation in Chemical Wet Deposition with Meteorological Conditions," *Atmos. Environ.* (in press).
6. Brezonik, P. L., E. S. Edgerton and C. D. Hendry. "Acid Precipitation and Sulfate Deposition in Florida," *Science* 208:1027–1029 (1980).
7. Caiazzia, R., K. D. Haye and J. D. Gallup. "Wet and Dry Deposition of Nutrients in Central Alberta," *Water Air Soil Poll.* 9:309–314 (1978).
8. Eaton, J. S., G. Likens and F. H. Bormann. "Wet and Dry Deposition of Sulfur at Hubbard Brook," in *Effects of Acid Precipitation on Terrestrial Ecosystems,* Proceedings of NATO Conference, Toronto, Canada, May 21–27, 1978, T. C. Hutchinson and M. Havas, Eds., Vol. 4, NATO Conference Series (New York: Plenum Press, 1978), pp. 69–75.
9. Eaton, J. S., G. Likens and F. H. Bormann. "The Input of Gaseous and Particulate Matter to a Forested Ecosystem," *Tellus* 3:546–551 (1978).
10. Likens, G. E., F. H. Bormann and J. S. Eaton. "Variations in Precipitation and Streamwater Chemistry at the Hubbard Brook Experimental Forest During 1964 to 1977," in *Effects of Acid Precipitation on Terrestrial Ecosystems,* Proceedings of NATO Conference, Toronto, Canada, May 21–27, 1978, T. C. Hutchinson and M. Havas, Eds., Vol. 4, NATO Conference Series (New York: Plenum Press, 1978), pp. 443–464.
11. Richardson, C. J., and G. E. Merva. "The Chemical Composition of Atmospheric Precipitation from Selected Stations in Michigan," *Water Air Soil Poll.* 6:385–393 (1976).
12. Schneider, W. A., W. R. Snyder and B. Clark. "Deposition of Nutrients and Major Ions in Precipitation in South Central Canada," *Water Air Soil Poll.* 12:171–185 (1979).
13. Tabatabai, M. A., and J. M. Laflen. "Nutrient Content of Precipitation over Iowa," *Water Air Soil Poll.* 6:361–373 (1976).
14. Raynor, G. S., and J. P. McNeil. "An Automatic Sequential Precipitation Sampler," *Atmos. Environ.* 13:149–155 (1979).

PART 3

THEORETICAL CONSIDERATIONS

CHAPTER 12

CRITIQUE OF METHODS TO MEASURE DRY DEPOSITION

Bruce B. Hicks* and Marvin L. Wesely

Argonne National Laboratory
Argonne, Illinois 60439

Jack L. Durham

U.S. Environmental Protection Agency
Research Triangle Park, North Carolina 27711

INTRODUCTION

Within the last decade, there have been improvements in urban air quality, but deterioration has occurred on the regional scale. Contributing factors are the shifting of electrical power generation to rural areas and use of tall stacks. There has been a recent awareness of the decline of visual quality in the eastern United States, and a serious concern for what now appears to be a significant trend of increasing quantities of strong acids removed from the atmosphere by both wet and dry deposition.

The subject of dry deposition addresses all processes by which airborne contaminants are removed from the atmosphere at the surface of the earth, excluding those processes directly aided by precipitation. Ultimately, knowledge of the fate of pollutants is desired, particularly if the effects are harmful. For surface vegetation, for example, dry deposition can sometimes be equated to a dose that causes specific harmful effects, but often the

---

*Present address: ATDL-NOAA, Oak Ridge, Tennessee 37830.

effects of pollutants on biological systems occur after transport away from the initial surface contacted and after some change in chemical form. Since the surface flux of atmospheric contaminants affects concentrations downwind, dry deposition affects the extent of areas in which significant concentrations of pollutants are found. Thus, dry deposition exerts a strong, indirect influence on concentrations of pollutants found in precipitation and on interactions between pollutants.

Of concern here is primarily the rate of removal by dry deposition of pollutants in the atmosphere, to some extent their immediate fate, and to a lesser extent, their effects, except those that aid in determining deposition rate. Methods used to assess or predict rates of dry deposition are extremely diverse, largely because the types of contaminants and surfaces vary greatly. Since the final assessment of dry deposition must take place for the actual environment considered, relevant studies have concentrated on experimental efforts in the field. Laboratory and purely theoretical work play a necessary role in understanding the processes that control dry deposition. Such understanding leads to a productive use of data from networks monitoring air quality, helps extrapolate present deposition results to pollutants not yet fully investigated, and leads to better parameterizations of surface removal for use in numerical models of pollutant distribution in the atmosphere. Presently, our knowledge of rates is too limited, and our ability to monitor in networks is severely restricted.

There has been a continuing interest in dry deposition for many years. Theory, methodology, and experimental results have been reviewed in an Energy Research and Development Agency symposium (1974) [1] and in many subsequent papers. However, it was not the intent of the U.S. Environmental Protection Agency (EPA) workshop on dry deposition (Dec 79) [2] to provide an updated summary of the status of this research field.

The purposes were: (1) to identify the advantages and disadvantages of current methods for measurement of dry deposition of air pollutants, (2) to speculate on the potential of other methods and (3) to make recommendations to the EPA for research and development in regard to (a) research methods, instruments, field measurements and tests, and (b) methods for evaluating and monitoring dry deposition on a routine or semiroutine basis.

The purposes were not: (1) to certify the methods and results obtained by the participants or others, or to lobby for a favorite method, (2) to address ecological effects, (3) to address transport modeling or (4) to obtain a consensus on everything.

In this chapter, we present a concise summary of the workshop's critique of methods, identification of research needs on processes, and the conclusions/recommendations. Not presented here, but contained in the complete workshop report, are: (1) brief descriptions, specific critiques and recommendations for individual methods that were discussed at the workshop, (2) list of participants and (3) the complete texts of the dissenting views.

CRITIQUE OF METHODS

This section presents brief descriptions and evaluations of various experimental methods for investigating and estimating dry deposition. The methods are representative of most research to date, and include some especially promising techniques. Purely theoretical efforts are not considered separately, but this practice is not meant to imply that such work is unimportant. In fact, all of the experimental methods are directed to some extent toward developing better theories on deposition; the parameterization studies to be considered are specifically concerned with theoretical and semiempirical expressions that affect deposition.

The main focus is on methods capable of determining dry-deposition rates of various pollutants either by measurement alone or by a combination of measurements with numerical simulations. The methods are sorted into three categories: (A) estimates of accumulation, (B) flux monitoring, and (C) flux parameterization. These distinctions are somewhat artificial; the categories only reflect the intent of the methods as they are currently reported in the scientific literature. A summary of the critiques is given in Table 1.

If the total accumulation of certain pollutants on a given area could be measured directly, the answers most desired might be found without further effort. This approach constitutes category A: estimates of accumulation. Periods of time ranging from months to years are usually considered; short-term variations cannot be identified. While areas ranging in size from the entire surface of the earth to individual roughness elements can be addressed in principle, a more practical size is that of individual watersheds characteristically from 1 to 100 km in horizontal dimension. The methods of estimating accumulation could potentially be used for routine monitoring, but the effort required would probably be too large.

Numerous attempts have been made to monitor dry deposition, i.e., to measure fluxes directly on a routine basis. These techniques usually involve sampling at individual points; they are considered in category B: flux monitoring. Single measurements typically last from hours to weeks, and so shorter-term variations are usually not detected. Ideally, the point measurements are taken over a well-defined surface, e.g., one of uniform vegetation. Actually, all of the point- and areal-sampling field methods listed in Table 1 are potential monitoring techniques, but those listed in category B are the ones that to date have been used most often for monitoring.

Table 1. Summary of Methods to Measure Dry Deposition
(Only those methods for particles that are not affected by gravitational settling are considered.)

Category	Method	Pollutant		Monitoring Pontential	Research Potential
		Particles (P)	Gases (G)		
A. Estimates of Accumulation	Atmospheric radioactivity	P		low[a]	medium[b]
	Mass-balance studies	P	G	low[a]	high[c]
B. Flux Monitoring	Open pots	P		low[d]	low[b,d]
	Flat filters	P		low[d]	low[b,d]
	Flat plates and shallow pans	P		low[d]	low[b,d,e]
	Fiber filters	P		low[d]	low[b,d]
	Sticky films	P		low[d]	low[b,d]
C. Flux Parameterization 1. Field work					
a. Large areas	Box-budget studies	P	G	low[a]	medium[f]
	Airborne eddy correlation	P	G	low[a]	high[g,h]
b. Small areas	Gradient	P	G	low[i]	high[j]
	Modified Bowen ratio	P	G	high	high[j,k,l]

Eddy correlation	P	G	low[i]	high[g]
Variance	P	G	high	high[e,j,k]
Tracer experiments	P	G	low[a]	medium[m]
Eddy accumulation	P	G	high	high[j]
Leaf washing	P	G	medium[a]	medium[n]
Surface snow sampling	P	G	medium[o]	medium[o]
2. Laboratory				
Chamber studies	P	G	low	high[p]
Wind-tunnel studies	P	G	low	high[q]

[a] Large effort required.

[b] Poor particle size discrimination

[c] Addresses fate of pollutants to some extent; does not discriminate between particles and gases.

[d] Easy to collect sample but difficult to interpret; contamination by gases possible.

[e] Alternative viewpoint expressed.

[f] Difficult to achieve the absolute accuracies required.

[g] Requires fast-response instrumentation.

[h] Only pollutant tried so far has been ozone.

[i] Instrumentation difficult to maintain.

[j] Susceptible to contamination by resuspension of particles.

[k] Not yet tried for pollutants.

[l] Requires small short-term drift of sensors.

[m] Data difficult to interpret.

[n] Leaf-to-leaf variability is large.

[o] Works only for light winds, subfreezing temperatures.

[p] Used mostly to identify important processes.

[q] Levels of turbulence found in the field difficult to achieve realistically.

The last, and largest, category of purpose to consider is C: flux parameterization. Many of the field methods listed may become useful for routine monitoring as technology develops; at present, however, the methods in category C, as opposed to those in category B, are used primarily for understanding and parameterizing deposition processes. With successful parameterizations, concentrations measured directly or computed in numerical models can be utilized to calculate the vertical fluxes. Category C is further divided into two subcategories according to whether the work is performed in the field or in the laboratory. For field work, both large areas of 1-100 km in horizontal extent and small areas of uniform surface characteristics are considered. Sampling times vary from 10 minutes to several days. During short-term studies over small areas, great effort usually must be made to obtain sufficient information on the physical, chemical and biological characteristics of the surface. For laboratory work, sampling periods are highly variable, but usually short. The type of laboratory observations considered here is concerned with bulk exchange processes in rather large enclosures.

INDIRECT CALCULATION OF DEPOSITION RATES

Another method for determining deposition is always implied to some extent when parameterization is discussed. This more indirect method is the application of parameterizations to measurements of pollutant concentrations and a few selected companion variables in order to compute the dry deposition rates indirectly. The use of concentrations monitored at a network of stations is emphasized here, but concentrations computed by numerical models are also inherently considered.

For airborne particulate and gaseous pollutants, there is no historical data base on dry deposition, except for some exceptional cases such as architectural surfaces. However, there are data bases on air quality and there are networks for measuring air quality in operation at the present time. Thus, an approach to monitoring dry deposition is to use the measurements of air-pollutant concentrations at a single height and employ parameterizations of dry deposition to yield the deposition rates. From micrometeorological observations and knowledge of surface behavior, deposition velocities are inferred. Pollutant flux is obtained as the product of deposition velocity and concentration. Companion variables that form the basis for the meteorological calculations are wind speed, atmospheric stability and surface roughness; depending on pollutant and surface characteristics, a range of other properties must be monitored. This method has an advantage in that monitoring stations are already in place for some pollutants. Unfortunately, however, single-height concentration monitoring is fundamentally deficient. Meteorological information alone will not yield a measurement of deposition velocity because it cannot account for all surface variables, such as the physiology of vegetation. Thus, careful observations of surface characteristics must be made. While it is probably not essential that the concentrations be measured at extremely uniform sites, micrometeorological measurements should be. The alternative, less exact but more practical method is to infer atmospheric variables from standard meteorological measurements and assume diurnal cycles of stability.

Until reliable methods suitable for directly monitoring dry deposition are developed, use of monitored concentrations may be the only alternative that is practical yet sufficiently accurate. Parameterizations of SO_2 and O_3 fluxes to many surfaces are already available, but are not well founded for submicron particles and many gases. For surfaces, additional work above tall plant canopies such as forests is needed.

SOME RESEARCH NEEDS

At attempt is made here to describe some of the processes controlling dry deposition that need to be better understood, and to suggest methods of research that might be employed. These considerations complement the flux parameterization methods identified in Table 1. The major problem in parameterizing the flux of specified pollution species to a given surface can often be identified by noting where the resistance to vertical transport is greatest. For example, uptakes of certain gases by vegetation is mainly controlled by the sizes of the leaf stomatal openings. For submicron particles, deposition to smooth inert surfaces is limited by the slow diffusion through the quasilaminar layer of air closest to the surface; any means of effectively short-circuiting that resistance can result in large changes in deposition rates.

Practical considerations limit the number of chemical species and types of surfaces that can be investigated. The surfaces chosen for study usually should be those most prevalent or characteristic of the area or region considered. At a minimum, the pollutants to be examined should be the criteria pollutants and particles in many small size intervals, with emphasis on submicron diameters. Of course, studies are aided by grouping together pollutants with similar chemical and physical properties. Categories might depend on solubility in water or reactivity with various substrates, for example. For particles, the processes of diffusion, impaction and sedimentation can sometimes be studied separately, but care must be taken because these are smooth functions of particle size in overlapping size ranges.

Table 2 summarizes identified research needs, grouped according to scales of the phenomena of interest and correlated with experimental methods of investigation given in Table 1. The smallest scale sizes, those associated with

Table 2. Summary of Research Presently Needed on Dry Deposition Processes

Contributing Process (resistance)	Length, Scale or Height (m)	Topics of Concern	Recommended Methods of Experimental Research
r_e, resistance of individual surface elements	≪1	Particles: surface wetness, stickiness, microscale roughness, electrostatic attraction Gases: numerous chemical and biological factors as they interact with pollutants having various properties[a]	All with medium or high research potential identified in Table 1
r_L, resistance of quasi-laminar layer enveloping surface elements	~0.005	Particle deposition as affected by impaction and diffusion[a]	Wind-tunnel studies plus all small-area field flux-parameterization techniques (C.1.b. in Table 1)
r_s, bulk surface resistance, composed mainly of r_e's and r_L's	<30	Relationship of mean vertical fluxes aloft to transport to individual surface elements	Micrometeorological techniques[b] applied in conjunction with leaf washing, tracer studies, mass-balance studies and similar methods
r_a, aerodynamic resistance in the atmospheric surface layer	~50, day ~5, night	Similarity theory applied to vertical gas transport[c]; transport in very stable conditions[c]	All micrometeorological techniques[b]
r_a and r_L as affected by medium-scale surface discontinuities	<30	Increased aerodynamic roughness, increased aeration of plants	Complex geometry[d]: box-budget studies, airborne eddy-correlation, atmospheric radioactivity, mass balance studies Simple geometry[d]: wind-tunnel studies, tracer studies

| r_b, resistance in the planetary boundary layer | ~1500, day ~100, night | Vertical flux divergence due to physical and chemical processes | Box-budget studies, airborne eddy correlation |

[a]Probably where initial research emphasis should be placed.
[b]The gradient, modified Bowen-ratio, eddy-correlation, variance, and eddy-accumulation methods.
[c]Not felt to be particularly important in research presently needed.
[d]Not specifically considered at the workshop.

r_e in Table 2, are often best studied in the laboratory, where, for example, reactivity with various substrates can be examined. Further work on particle deposition is needed to determine the effects of surface stickiness and micro-scale roughness. Continued work on diffusiophoresis, thermophoresis and electrophoresis is needed, but with greater emphasis on the environmental conditions that exist in the field. For example, electrical charges on particles vary with the "age" of the aerosol, resulting in varying effects of electro-phoresis on deposition. Another problem is that pollutants often are (re)sus-pended or (re)emitted, due to a variety of causes that may not be appreciated until what appears as anomolous results are found in laboratory and field experiments. Small-scale surface phenomena that need to be studied in this respect include decomposition of organic material and chemical reactions at surfaces that "recycle" pollutants.

Next to the actual surface is the quasilaminar boundary layer with a resis-tance r_L often considered separately. A considerable amount of wind-tunnel work has been performed on phenomena of this scale size. If possible, both laboratory and theoretical studies should consider the high levels of atmo-spheric turbulence and its effects on the interfacial air layer. For example, the intermittent penetration of strong gusts into deep plant canopies during unstable conditions strongly modifies both the concentration of certain pollutants and the thickness of the boundary layers of air around surface elements.

A difficult theoretical and experimental problem concerns the process by which transfer to surface elements is added up to yield the total deposition from the atmosphere. Micrometeorological techniques usually only yield an estimate of bulk surface resistance (r_s in Table 2) that may not be adequate by itself to identify the controlling process. For present purposes, the goal sought should be simple parameterizations of bulk surface and interfacial sublayer resistances or filtration efficiency. Both numerical and analytical models have been attempted; they usually rely to some extent on empirical relationships found in laboratory and field work. A by-product of such models is the calculation of the accumulation of harmful pollutants to parts of the surface.

On the largest scale sizes, those concerning r_a and r_b in Table 2, applica-tion of the parameterizations gives deposition velocities for uniform sur-faces; the deposition rates for large diverse areas are computed as an average of contributions from each surface weighted in proportion to the total area covered by each surface. Several sources of error can affect the estimates. For example, the fluxes themselves alter the concentrations, so that use of a single concentration for many types of surfaces is inaccurate. Also, the effects of surface discontinuities may result in altered deposition rates near boundaries of an area with a single type of surface. Obstacles such as trees and isolated buildings have an effect, not to mention the unknown extent to which complex terrain can alter the parameterizations. Obviously, experimental and theoretical methods used to assess these effects should be developed further. Perhaps box-budget experiments can be used to determine errors; such procedures may in the future be sensitive enough to detect the large errors that are of greatest concern.

CONCLUSIONS AND RECOMMENDATIONS

The rate of dry deposition of atmospheric pollutants depends on many factors concerning the pollutants, surfaces and atmospheric conditions. To evaluate methods that might potentially be used to monitor dry deposition, the roles of all contributing processes must be considered, at least in broad categories. The most important processes that affect depositon rates can often be identified as those corresponding to the largest resistance to vertical transport or uptake.

For small particles that are deposited by diffusion and impaction, deposition is strongly affected by the resistance of the quasilaminar sublayer of air nearest to surface elements and by the microscale roughness and stickiness of the surface. For gaseous pollutants, the chemical and biological nature of a surface, as well as its physical configuration, often dominates the resistance to air-surface exchange. Processes of (re)emission and (re)suspension, which can greatly affect net deposition, are also highly dependent on surface properties. Hence, the following recommendation can be made:

1. Since use of artificial collecting surfaces does not simulate the net dry deposition of particles and gases to natural surfaces, it would be ill-advised to rely upon continued monitoring by "open pot" or various other types of surrogate surfaces. As a result, present capabilities to monitor dry deposition in a practical, yet accurate, manner are inadequate.

Use of such surfaces has continued for estimating deposition of particles in some cases, simply because of the lack of better procedures that can be utilized with small cost and effort. Notably, knowledge on how to "calibrate" surrogate surfaces for gaseous or particulate deposition will probably never be obtained. At best, future efforts might relate the amount and type of material collected by a surrogate surface to a corresponding atmospheric concentration; it would then remain necessary to deduce surface fluxes from the inferred concentrations, not an easy task in most cases at present. It would probably be easier and more accurate over natural surfaces to measure concentrations directly rather than relying on artificial surfaces intended to simulate the actual surface.

In the process of choosing a site for monitoring (which assumes that a suitable monitoring technique is available), the importance of knowledge of surface characteristics must be emphasized. This consideration is reflected in another recommendation:

> 2. Sites chosen for monitoring should have a surface "typical" of the surrounding area; in the immediate vicinity of measurements, the surface should be a single uniform type. The condition of local vegetation and the general environmental conditions that affect the surface should be recorded.

Obviously, there might be some difficulty in finding suitable sites in many locations. An additional requirement for site selection is:

> 3. Monitoring sites should have pollutant and wind characteristics representative of the large surrounding area. Records should be maintained on the possible presence of pollutants from local sources and on local meteorological events such as precipitation, fog and dewfall.

The above two requirements for a good site imply that sites currently chosen for wet and dry deposition monitoring may not be adequate for obtaining representative estimates of dry deposition.

For monitoring, estimates of atmospheric resistance should be made routinely, since this resistance can control the transfer of pollutants that can be removed very efficiently by the surface, and since it is important to identify conditions of atmospheric decoupling at night, when turbulent vertical transport becomes exceedingly small. Provided a few essential measurements of atmospheric conditions are taken, micrometeorological formulations currently available are adequate for estimating the atmospheric resistance to vertical transport from a height of several meters to fairly close to the surface. It is recommended that:

> 4. At each monitoring site, wind speed, atmospheric stability, and aerodynamic surface roughness should be determined, so that atmospheric resistances can be calculated and averaged over time intervals preferably not to exceed six hours; one-half to one hour is optimum.

Current techniques intended to monitor dry deposition directly on a routine basis do not appear to be adequate. As stated above, surrogate collecting surfaces do not simulate surface conditions sufficiently well in most cases. An alternative method is to attempt to infer surface fluxes from pollutant concentrations measured at a single height; much work has already been devoted to develop the technical means to perform the needed measurements. Concentration measurements at one height, however, give no information on transfer processes in the atmosphere and at the air-surface interface. The following recommendation must therefore be made:

5a. Concentration measurement at a single height should not be used by itself to provide a measure of dry deposition.

The availability of measurements of some pollutant concentrations leads to questions as to what supporting information is required in order to interpret these measurements. Deposition rates can be estimated as the product of concentration and an appropriate deposition velocity. The latter quantity can be derived from accurate parameterizations of the deposition processes. As implied in Recommendation 4 above, micrometeorological aspects of parameterizations can be estimated fairly well above uniform sites. However, interactions of pollutants with the surfaces and the behavior of small particles in the interfacial air layers are not well understood. Therefore, to interpret past and future measurements on concentrations, at least until means sufficient for accurately and routinely monitoring dry deposition are developed, further research is needed. The following statement concludes Recommendation 5:

5b. In order to understand the relationships between vertical fluxes and concentrations measured at a single point, research on the parameterization of deposition processes should continue by use of a variety of theoretical approaches and experimental techniques in the laboratory and the field.

Cooperative efforts between micrometeorologists, biologists and aerosol scientists, among others, are required; many recent studies have not utilized such cooperation. For example, atmospheric investigations often do not study a minimum set of variables concerning pollutant and surface characteristics in order to elucidate the deposition processes. On the other hand, ecological studies often do not adequately document atmospheric conditions.

Based on present knowledge, it is possible to identify certain information that would need to be obtained routinely at monitoring sites in order to estimate deposition velocities for evaluating surface fluxes from concentrations measured at a single height. Of course, this needed information must also be obtained during the parameterization studies that form the basis for this possible monitoring technique. As already stated, wind speed at some specified height above the surface should be measured and atmospheric stability estimated, both for periods not to exceed six hours. The type, height and condition of vegetation present should be known. Furthermore, consideration should be given to a number of other factors that appear to be important in the case of certain pollutant species. Soil moisture content can affect the uptake of some gaseous pollutants, especially over bare soil. Some measure of aridity and atmospheric stability might suffice to describe the gross aspects of physical environment that affect plant physiological functions; relative humidity and solar radiation measurements might suit this purpose. Other variables, particularly those dealing with conditions of vegetation, probably will be identified as important yet suitable for routine evaluation as a result of future parameterization studies.

Deposition velocities depend on the type of pollutant as well as on surface and atmospheric conditions. For particles, studies have indicated that deposition is affected by diffusion, impaction and sedimentation processes that are continuous functions of particle size in overlapping ranges. Measurements employing a single size fractionation of particles (and hence providing samples of "large" and "small" particles) are not satisfactory, since it is likely that the coarse and fine fractions each must be resolved into several size classes. Gases can be considered in categories of similar chemical properties only to a limited extent. Thus, to infer dry deposition from measurements of concentration at a single height, direct measurement of each pollutant of interest is required. The possible importance of (re)suspension and (re)emission processes further underlies this need.

To perform parameterization research such as implied in Recommendation 5b, one would surmise that there are or will be available methods to measure reliably the deposition rates. In that case, the measurement techniques might themselves offer the best means to monitor deposition rates. The corresponding recommendation is:

6. Techniques used in field parameterization studies should be developed and examined for applicability to routine monitoring.

The methods most recommended are micrometeorological in nature. Verification of the techniques requires some direct measurements of accumulation at the surface; box-budget experiments, though usually too inaccurate, may someday provide an estimate of errors. The following statement, at least, can be made:

7. Although micrometeorological methods offer the promise of providing a direct and precise measure of dry deposition, experimental complexity and sensor inadequacies combine to reduce confidence in them. At this time, no micrometeorological method can be accepted with absolute confidence. There is need for continued intercomparison and comparison with other methods.

Fluxes of particles large enough to be affected significantly by gravitational settling cannot be measured accurately by standard micrometeorological techniques. In fact, if such large particles are not excluded from detection in some techniques, erroneous results might be obtained.

Some micrometeorological methods can be identified as possibly worthy for development for use in monitoring. They are the variance method, the eddy-accumulation techniques, and the modified Bowen-ratio technique. Others may be indicated as research progresses. For the variance methods, the main limitation seems to be in the ability to find or develop chemical sensors that have fast response (at least 1 Hz) in addition to being linear and having low noise. The other two methods, and others likely to be suggested, rely on accurate mean measurements of concentration differences. Therefore, the following recommendation is made:

8. Methods to measure dry deposition that depend on concentration differences detected in samples collected in the atmosphere must be supported by chemical/physical analysis methods with standard errors of 1% or less.

This recommendation does not necessarily imply that absolute accuracies must be 1% or less. If the modified Bowen-ratio technique is used, for example, a single sensor might be rigged to alternately measure at two heights. Then the requirement is that the difference, which will usually be 1–10% of the mean, be measured accurately. Thus sensor offsets and slow drifts have almost no effect, and errors, say of $\sim 10\%$, in span calibrations would produce errors of only $\sim 10\%$ in the estimated vertical flux.

Techniques to measure, let alone monitor, fluxes of species associated with acid deposition are presently deficient. Many of the reasons have already been given in this section. There is an urgent need to develop methods with the required accuracy in measuring amounts of strong acids and species such as H_2SO_4, HNO_3, SO_4^{2-}, NO_3^-, NH_4^+, NH_3, and organic acids and bases. This need leads to the following recommendation:

9. Present capabilities to measure dry deposition for parameterization studies as well as for monitoring must be improved in order to support the acid deposition programs of EPA.

DISSENTING VIEWS

Potential of Micrometeorological Methods

Dry deposition measurement methodology is a combination of two disciplines (1) the micrometeorological measurement or inference of air transport parameters and (2) quantitative chemical analysis. The workshop has critiqued various methods mainly from the viewpoint of physical mass transport parameters, which is expected since most of the workshop participants were micrometeorologists. The degree of accuracy of chemical analysis is stated for several of the techniques; however, the unreasonable difficulty of such demands on the chemical analyst is not discussed. Not demonstrated is the requirement of nonsurrogate surface dry deposition measurement procedures to be able (a) to detect mean concentration differences of 1–15% in successive samples to within an accuracy of 10%, or (b) to monitor pollutant air concentrations with a response time of one second or less. However, if such requirements are valid, the successful achievement of chemical/physical measurements in a routine monitoring network is an unreasonable expectation. In general the error in flowrate measurement for aspirated samples is not better than ±10%. If this fluctuation error could be reduced to ±1% (which is unlikely), the chemical measurement random errors are expected to be too large to be acceptable.

The extreme requirements for precision/response time of the chemical analysis/monitor make it unlikely that such methods will be developed rapidly, as evidenced by the total absence of even one routine dry deposition monitor based on eddy accumulation, modified Bowen-ratio, or variance techniques.

Sample Integrity

While a legitimate need may exist to monitor the "dry depositon of pollutants, especially the strong mineral acids," known atmospheric chemical behavior is ignored in regard to the preservation of sample integrity. For example, the presence of strong mineral acids in fine aerosols has been demonstrated; however, the presence of such acids in a dry deposition *sample* (from a bucket, plate or filter collector) is highly improbable because eventual neutralization by ambient ammonia will occur over the collection period (several days). Of at least equal importance is the simultaneous collection of large mineral particles that are bases, which neutralize the acids during extraction. Our previous experience has been that dust-fall bucket samples invariably yield a *basic* aqueous extract. Such a result is contrary to the suspected acidification of surface waters in the northeast.

Use of Surrogate Surfaces

Surrogate surface depositon measurements of various types have been conducted for years, and will probably continue to be used on a widespread basis. Simplicity in obtaining the data, control over contamination, ability to collect sufficient material for trace level analysis, and low expenses are the primary reasons. It is acknowledged that certain applications of such measurements in particular may not provide as much accuracy as more sophisticated techniques; however sufficient data are available to establish conclusively that all types of surrogate surfaces should not be used.

Several surrogate surface studies reported in the literature have shown consistencies in the data which suggest that semiquantitative information can be obtained. (The results of eight separate investigations are cited in the workshop report [2].)

Recommended research includes determination of the relationships among material suspended in the atmosphere, collected by surrogate surfaces and deposited on natural surface elements (as determined by washing and microscopic examination, or other techniques). Such comparison studies might lead to a better design and application of certain artificial surfaces than those routinely used for monitoring, such as the use of flat plates situated in the foliar canopy as opposed to the use of buckets at ground level.

SUMMARY

The workshop report [2] provides evaluations of various methods for estimating dry deposition of gaseous and particulate pollutants from the atmosphere to the surface of the earth. The intent is to provide the scientific information needed in order that sound decisions can be made on research development and coordination with regard to dry deposition. Existing and potential techniques for monitoring and parameterizing dry deposition are considered systematically and specific recommendations are given for each method on its present and potential usefulness. The viewpoints presented are based on and limited to opinions expressed during discussion that took place at the workshop. A priority of research needs was not established at the workshop and thus is not given here.

The current capabilities for monitoring dry deposition are felt to be inadequate because the methods now used, which usually employ surrogate surfaces or open vessels, do not give the deposition at the actual natural surface. Most of the workshop participants felt that "calibration" of such methods would not be successful or have very limited success. However, 20-30% of the scientists present who have been or are making original contributions to scientific research on dry deposition feel that the use of surrogate surfaces should be pursued in some fashion.

Fairly direct measurements of the net dry deposition can be obtained by measurement of the accumulation of certain chemicals in rather large areas over long periods of time. Because of the considerable amounts of effort necessary in application, this category of method clearly was never intended to apply to routine monitoring, although the results of such studies are quite useful in determining the fate of pollutants after reaching the surface.

Parameterization studies are directed toward understanding the relationships between vertical fluxes and the properties of pollutants, surfaces and the atmosphere. Estimation of deposition rates needed for numerical simulations of pollutant behavior in the lower atmosphere usually are based on parameterizations of the deposition velocity (the ratio of flux to concentration at a specified height in the atmosphere). In the report, evaluations of numerous methods are presented. Of course, knowledge accumulated from parameterization studies might allow the use of pollutant concentrations monitored at a single height to estimate the vertical flux. To date, very few species of pollutants have been studied adequately to approach this goal. For example, the behavior of many species associated with strong acids has not been determined, because the chemical sensors presently available do not meet the specifications required by the techniques for measuring dry deposition. In many cases, there is no "sensor" but instead a procedure for chemical analysis best performed in a well-equipped laboratory. Usually, measurement procedures should be able to detect mean differences of 1–10% in successive samples to within an accuracy of 10%, in order to be useful to measure deposition rates. For flux measurement techniques that rely on detection of turbulent fluctuations in pollutant concentrations near the ground, the speed of response shoud be one second or less.

A simple, but accurate, method for monitoring dry deposition rates routinely is not available. Micrometeorological methods that have potential in this regard are the techniques of eddy accumulation, modified Bowen-ratio and variance methods. It should be remembered, though, that even if these or other methods prove very useful, the requirements on the selectivity and performance of chemical and physical sensors are still quite high and, in many cases, not yet achievable.

Sites selected for parameterization and monitoring studies ideally should be located in uniform terrain so that results are not affected by unusual features of the surface unique to that location. Monitoring sites, if they are to be effective, should be located at surfaces that are typical of the surrounding areas and that should have representative meteorological conditions and pollutant characteristics.

ACKNOWLEDGMENT

The U.S. Environmental Protection Agency thanks the workshop participants for their contributions and Argonne National Laboratory, Argonne, Illinois, for hosting the meeting.

REFERENCES

1. "Atmosphere-Surface Exchange of Particulate and Gaseous Pollutants (1974)," Symposium Proceedings, Richland, WA, 4-6 Sep 74. U.S. Energy Research and Development Administration Report No. CONF-740921 (1976), available from NTIS.
2. Hicks, B. B., M. L. Wesely and J. L. Durham. "Critique of Methods to Measure Dry Deposition," U.S. EPA Report EPA-600/9-80-050 (1980), available from NTIS (No. PB81 126443).

CHAPTER 13

AN INTEGRATED APPROACH TO
ACID RAINFALL ASSESSMENTS

J. L. Schnoor, G. R. Carmichael and F. A. Van Schepen

Engineering College
The University of Iowa
Iowa City, Iowa 52242

INTRODUCTION

A tiered approach to the mathematical modeling of acid precipitation, its fate and effects, might include: (1) a steady-state susceptibility model to assess long-term effects from average annual atmospheric loadings on surface and groundwaters; (2) a dynamic, regional model to assess the role of long-range versus point-source pollutant loadings for regulatory purposes and (3) a three-dimensional dynamic event model to assess the maximum environmental insult to a specific ecosystem during critical conditions such as a snow-melt event. An integrated approach is needed which considers the fate of pollutants from their origin through terrestrial and aquatic ecosystems to the ultimate water resource, groundwater. In this chapter, the first step, the steady-state model is developed and applied to northern Wisconsin, the Upper Wisconsin River Basin. An example of the dynamic problem including CO_2 gas exchange and equilibrium pH is also discussed.

TRICKLE-DOWN MODEL

Conceptual Development

Figure 1 is a schematic of a compartmentalized fate and transport model for acid precipitation assessments. The driving force for the system is the natural and anthropogenic emissions of SO_x and NO_x. Other chemical emissions such as H_2S and N_2O may be oxidized relatively rapidly in the atmosphere. The model is a "trickle-down" system in which pollutants descend from the atmosphere to the terrestrial canopy where sorption and biotransformations occur. The terrestrial compartment includes the snow-pack (and its attendant deposition) as well as marshes and overland runoff water. Rivers and lakes are classified as surface waters. Terrestrial runoff water becomes tributary to lakes, and percolation results in surficial or water table aquifers. There may exist interflow between the surface water and surficial aquifers, or alternatively groundwater recharge of bedrock aquifers. Acid is neutralized by dissolution of calcerous overburden and other minerals, and leaching of nutrients or heavy metals is possible along the way.

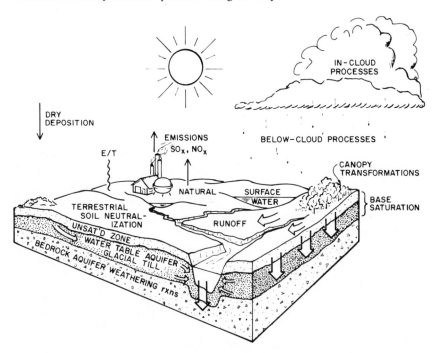

Figure 1. Schematic of compartmentalized fate and transport model for acid precipitation assessments. Compartments include clouds, below-clouds, canopy, soil layer, unsaturated zone, water table aquifer, surface lakes and streams, and bedrock aquifers.

Figure 2 depicts the transport of atmospheric pollutants through a number of interconnected compartments in the grid. Convection and dispersion connects the well-mixed boxes of volume, $V = \partial x \cdot \partial y \cdot H$, where ∂x and ∂y are the length and width of the cell, respectively, and H is the average mixing-layer height. Initially, the surface flux calculations are based on cell con-

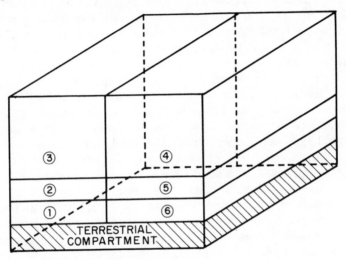

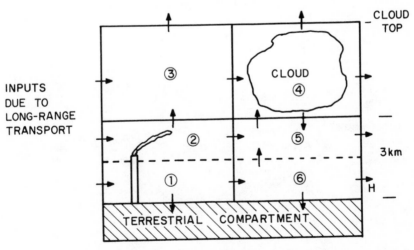

Figure 2. Interrelationship of atmospheric transport model and terrestrial compartment.

centrations determined from actual field data. Twenty-four-hour averaged SO_2, $SO_4{}^{2-}$, $NO_3{}^-$ and TSP concentrations measured at monitoring stations in the vicinity of the study region have been compiled from existing data bases (e.g., EPRI/SURE, MAP3S, etc.) and are used to estimate 24-hr averaged concentrations within each cell. These concentrations are then used to calculate the amounts of SO_2, $SO_4{}^{2-}$ and $NO_3{}^-$ deposited to the surface in that 24-hr period. The dry deposition velocities of each cell are determined based on surface type and season, and the wet depositon rate within each cell varies with precipitation amount and duration. Seasonal and yearly loadings are then estimated as steady-state approximations. This analysis can be extended to other trace gases (e.g., NH_3).

To develop an accurate mass balance model for acidity transfers between the five compartments in the trickle-down model requires a good understanding of the geohydrology within each cell. Figure 3 presents a hypothetical hydrologic balance for the Rhinelander cell of a northern Wisconsin acid rainfall assessment. Assuming steady state, the sum of the inflows to each compartment must equal the sum of the outflows plus evapotranspiration. Annual rainfall and evaporation estimates can be obtained from raingauge and pan information: terrestrial runoff and percolation are derived from the rational formula: runoff = rainfall intensity x area x fraction percolated. Aquifer recharge and export (out of the cell) can be calculated from knowledge of the soils, confining layers and bedrock geology via the 2-D Darcy equation. Permeability constants, porosities, piezometric head and

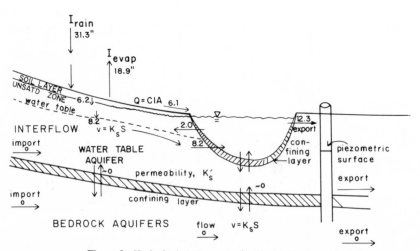

Figure 3. Hydrologic balance (in./yr) in trickle-down model.

aquifer dimensions are required for analysis. The hydrology of the system is critical to understanding both the steady state and the time response of the system.

Mass Balance Equations

The trickle-down model is based on the principle of mass continuity for alkalinity. Alkalinity is a conservative state variable which is not influenced by the exchange of CO_2 with the atmosphere. Negative alkalinity is free mineral acidity.

For each of the five compartments in the trickle-down model, an alkalinity mass balance equation is constructed according to Equation 1:

$$V_i \frac{dA_i}{dt} = \sum_{j=1}^{5} (Q_{i,j} A_j) - \sum_{j=1}^{5} (Q_{j,i} A_i) \pm \text{terms} \qquad (1)$$

$$\begin{array}{ccccccc} \text{Mass} & & \text{Sum of the} & & \text{Sum of the} & & \text{Sources} \\ \text{Accumulation} & = & \text{Inputs} & - & \text{Outflows} & \pm & \text{or Sinks} \end{array}$$

where

V_i = volume of the ith compartment (m^3)
$Q_{i,j}$ = flow from the jth compartment into the ith compartment (m^3/yr)
A_i = alkalinity of the ith compartment (eq/m^3)
t = time (yr)

As a first approximation for northern Wisconsin, the pH and pCO_2 of the rainfall were known, which specifies the alkalinity and equilibrium chemistry for the system $(CO_2, CaCO_3, \text{and } H_2O)$ according to Equation 2:

$$A = K_H (\alpha_1 + 2\alpha_2)(pCO_2/\alpha_0) + OH^- - H^+ \qquad (2)$$

where

A = alkalinity (eq/m^3)
K_H = Henry's law constant for CO_2
α_0 = $[H_2CO_3]/C_T$ = distribution coefficient
α_1 = $[HCO_3^-]/C_T$ = distribution coefficient
α_2 = $[CO_3^{-2}]/C_T$ = distribution coefficient
pCO_2 = partial pressure of CO_2 at 1 atmosphere = $10^{-3.5}$

Therefore, the system of five equations representing five compartments can be reduced to a system of four equations for four compartments and a specified wetfall and dryfall input. The average total sulfate load in the

Boundary Waters Canoe Area of Northern Minnesota has been reported as 11 kg/ha-yr by Glass and Loucks [1], and this figure has been used as an initial estimate of the total acid deposition. Best results were obtained with a sulfate loading of 24 kg/ha-yr, typical of northern Wisconsin.

The final system of mass balances for each compartment is presented in Equations 3-6.

Soil Compartment 2
$$V_2 \frac{dA_2}{dt} = Q_{2,1}A_1 + LA_{dry} \,(\% \text{ terrestrial}) - Q_{4,2}A_2 - \quad (3)$$

$$Q_{3,2}A_2 + k_2 M_2 V_2 (r_{eq_2} - r_2) - Q_{1,2}A_2$$

Unsaturated
Compartment 3
$$V_3 \frac{dA_3}{dt} = Q_{3,2}A_2 - Q_{5,3}A_3 + Q_{3,4}A_4 + k_3 M_3 V_3 \,(r_{eq_3} - r_3) \quad (4)$$

Surface Lakes
Compartment 4
$$V_4 \frac{dA_4}{dt} = Q_{4,2}A_2 + Q_{4,5}A_5 - Q_{6,4}A_4 - Q_{3,4}A_4 + Q_{4,1}A_1 \quad (5)$$

$$+ LA_{dry} \,(\% \text{ lakes})$$

Water Table Aquifer
Compartment 5
$$V_5 \frac{dA_5}{dt} = Q_{5,3}A_3 - Q_{4,5}A_5 + k_5 M_5 V_5 \,(r_{eq_5} - r_5) \quad (6)$$

Bedrock Aquifer
Compartment 6
$$V_6 \frac{dA_6}{dt} = Q_{6,5}A_5 - Q_{5,6}A_6 - Q_{7,6}A_6$$

where

L = dryfall acid loading (kg/yr as H^+)
A_{dry} = dry deposition (eq/kg as H^+)
k = rate constant for acid neutralization by soil (per year)
M = specific densty of soil in the overburden (kg/m^3)
V = volume of the interflow compartment (m^3)
r_{eq} = initial sum of the bases on the soil (eq/100 kg)
r = cumulative base cation export or equivalents of acid neutralized by soil (eq/100 kg).

Compartment 1 is the atmosphere and compartment 7 refers to export out of the grid cell. In this preliminary analysis, imports and exports of acid from the subsurface compartments to other grid cells were assumed negligible. The soil layer is usually a few inches to a foot deep. The unsaturated zone is primarily sand and gravel glacial deposit with some glacial till and is an average of 25 ft deep. The water table aquifer is 100 ft of sand and gravel, and the bedrock is crystalline rock (mostly granite, gneiss) ~ 1000 ft thick.

The acid neutralizing capability of the deep aquifer was neglected relative to the weathering of overburden in the soil and unsaturated compartments. The direction of annual average flow was taken to be from the interflow to the deep aquifer, from the interflow to the surface lakes, and from the surface lakes to the deep aquifers. Further geohydrological studies are needed to confirm this pattern [2].

Equations 3, 4 and 6 contain a term for sources of alkalinity (acid neutralizing capacity) in the overburden and soils, k M V (r_{eq} - r). This term provides a source of alkalinity until the soil's neutralizing capacity or base saturation is exhausted, i.e., r_{eq} = r. Even the surface water sediments can have a significant degree of buffering capacity under certain circumstances [3], but this factor was not included in the preliminary analysis. Soils data were available from the Soil Conservation Service on soil depth, bulk density, sum of the bases, porosity and hydraulic conductivity. The sum of the bases (or base saturation) was utilized as the value for r_{eq} in each soil layer. The other soils information was utilized in estimating the hydrologic balance and residence times for each compartment.

A more appropriate kinetic argument for the buffering capacity of the solid phase should reflect the mechanisms of H^+ sorption, cation exchange and incongruent dissolution. One possible kinetic expression which would demonstrate a gradual exhaustion or breakthrough curve for the solid phase is:

$$\frac{dr}{dt} = k_f[H^+] \; (r_{max} - r) - k_r[B] \, r \qquad (7)$$

where
r = equivalents of H^+ sorbed per mass of soil (eq/100 kg)
r_{max} = maximum base cation exchange capacity for H^+ (eq/100 kg)
k_f = rate of H^+ sorption (m^3/eq-yr)
k_r = rate of H^+ exchange with other cations (m^3/eq-yr)
B = export of base cations (eq/m^3)

Steady-State Results

Under steady state conditions, Equations 3-6 reduce to a set of four algebraic equations which are quite easily solvable by substitution. This model was applied to the Upper Wisconsin River Basin, a 2730-mi^2 area near Rhinelander, Wisconsin (Figure 4). Rainfall pH in the vicinity has been

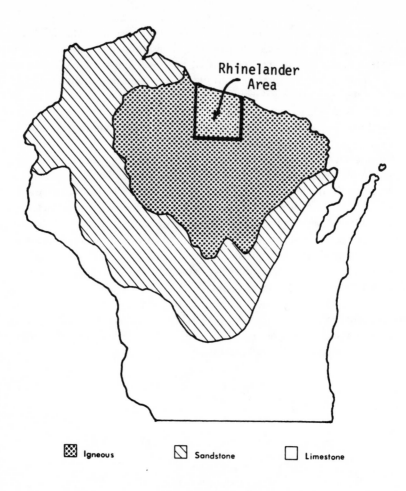

Figure 4. Rhinelander area site map.

averaging ~4.4 (Figure 5). The best estimate at coefficients for this cell using the data from Oakes and Cotter [2] and the SCS are:

V_2 = 1.55 x 10^8 m^3 soil water
V_3 = 2.59 x 10^9 m^3 unsaturated zone water
V_4 = 8.62 x 10^9 m^3 surface lakes
V_5 = 6.02 x 10^{10} m^3 water table aquifer water
$Q_{2,1}$ = 4.50 x 10^9 m^3/yr   rainfall on land
$Q_{4,2}$ = 8.95 x 10^8 m^3/yr   runoff on lakes
$Q_{3,2}$ = 8.95 x 10^8 m^3/yr   percolation to unsaturated zone
$Q_{5,3}$ = 1.25 x 10^9 m^3/yr   recharge to water table

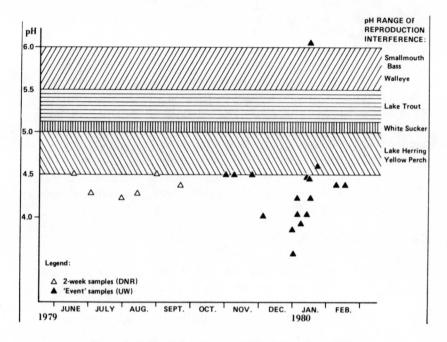

Figure 5. Acid precipitation near Rhinelander (from [4]).

$Q_{3,4}$	$= 3.59 \times 10^8 \ m^3/yr$	surface lakes to unsaturated zone
$Q_{4,5}$	$= 1.25 \times 10^9 \ m^3/yr$	water table to lakes
$Q_{6,4}$	$= 2.23 \times 10^9 \ m^3/yr$	export from surface lakes
$Q_{4,1}$	$= 1.12 \times 10^9 \ m^3/yr$	rainfall on lakes
L_{dry}	$= 1.42 \times 10^5 \ kg/yr \ H^+$	dryfall on land
L_{dry}	$= 3.55 \times 10^4 \ kg/yr \ H^+$	dryfall on lakes
A_1	$= -3.15 \times 10^{-2} eq/m^3$	rainfall alkalinity (pH 4.5)
A_{dry}	$= -1.0 \times 10^3 \ eq/kg$	dryfall alkalinity
M_2	$= 1.35 \times 10^3 \ kg/m^3$	specific soil density
M_3	$= 1.40 \times 10^3 \ kg/m^3$	specific soil density
M_5	$= 1.55 \times 10^3 \ kg/m^3$	specific soil density
β_2	$= 0.09 \ in./in.$	water capacity in soil
β_3	$= 0.06 \ in./in.$	water capacity in unsaturated zone
r_{eq_2}	$= 5 \ eq/100 \ kg$	sum of the bases in soil
r_{eq_3}	$= 0.6 \ eq/100 \ kg$	sum of the bases in unsaturated sands
r_{eq_5}	$= 0.1 \ eq/100 \ kg$	sum of the base in water table sands

Once the alkalinity balance is solved, it is necessary to solve Equation 2 with a root-finding technique in order to determine the pH of each compartment. The partial pressure of CO_2 in the aquifers was assumed to be three times greater than atmospheric in the interflow and ten times greater than atmospheric in the bedrock aquifer compartment.

Results are presented in Table 1, a sensitivity analysis on the weathering rates (represented as the zero-order reactions in Equations 3-6). For the steady-state analysis, it was assumed that the rates of weathering were constant, i.e., the base saturation utilized (r) was not allowed to vary over time. This amounts to an assumption that bases are produced by weathering reactions at the same rate that they are consumed in soil processes. In time-variable simulations, the equivalents of acid neutralized by the soil in weathering reactions is allowed to reach its saturation level r_{eq}. Steady-state results indicate the equilibrium pH values in each compartment for various assumptions on weathering rates. Case 4 is the most realistic according to present field observations in the Upper Wisconsin River Basin. Weathering produces a total of ~ 1200 eq H^+/ha-yr.

Time Variable Results

Equations 3-6 were also solved in a time variable simulation to determine the response of the system and the time required to 95% of equilibrium. The time to equilibrium is strictly a function of the hydraulic detention time and the rate of acid neutralization. The detention times were calculated as:

$$
\begin{aligned}
\text{soil} &= 12.6 \text{ days} \\
\text{unsaturated zone} &= 2.07 \text{ years} \\
\text{lakes} &= 2.64 \text{ years} \\
\text{water table} &= 48.2 \text{ years}
\end{aligned}
$$

The time required to achieve 95% of the steady state alkalinity after an impulse or a step function change is three times the mean hydraulic detention time of the compartment, but it can be lengthened considerably by the weathering reactions in Equations 3-6.

EVENT MODEL

An event model is needed to simulate the worst case in a lake or stream, such as a snowmelt or rainfall event during fish spawning. Since pH is the variable causing the biological effect, a time-variable pH model is required. Total alkalinity, CO_2, pH and CO_2-acidity are all interrelated. In an open system, total inorganic carbon changes due to gaseous equilibrium of CO_2 (Henry's law). If one knows two of the above chemical concentrations, it is possible to calculate all others at chemical equilibrium. The following is a diffuse source model for acid loading and titration of alkalinity in an open system.

Table 1. Sensitivity Analyses: Changes in Weathering Rates[a]

	Case 1	Case 2	Case 3	Case 4
Weathering Rates	$k_2 = 0$ $k_3 = 0$ $k_5 = 0$	$k_2 = 0.03$ $k_3 = 0$ $k_5 = 0$	$k_2 = 0.03$ $k_3 = 0.001$ $k_5 = 0$	$k_2 = 0.0254/yr$ $k_3 = 0.0005/yr$ $k_5 = 0.0038/yr$
Soil Compartment	pH = 3.8 Alk = −0.158	pH = 6.1 Alk = +0.0169 Wthr = 554	pH = 6.1 Alk = +0.0169 Wthr = 554	pH = 5.0 Alk = −0.01 eq/m^3 Wthr = 469 eqH$^+$/ha/yr
Unsaturated Zone Compartment	pH = 3.8 Alk = −0.158	pH = 5.8 Alk = +0.007	pH = 6.4 Alk = +0.028 Wthr = 38	pH = 6.5 Alk = +0.0372 eq/m^3 Wthr = 19 eqH$^+$/ha-yr
Surface Water Compartment	pH = 3.8 Alk = −0.158	pH = 4.7 Alk = −0.018	pH = 5.1 Alk = −0.007	pH = 7.5 Alk = +0.124 eq/m^3
Water Table Compartment	pH = 3.8 Alk = −0.158	pH = 5.4 Alk = +0.007	pH = 5.9 Alk = +0.028	pH = 6.9 Alk = +0.320 eq/m^3 Wthr = 626 eqH$^+$/ha-yr

[a]Assumed: dry deposition = 5.5 kg/ha·yr $SO_4^=$ = 177 eq/yr H$^+$.

Consider the kinetic equations for total inorganic carbon

$$\frac{dC_T}{dt} = K_a ([CO_2]_s - [CO_2]) + S_{c_T} \tag{10}$$

and for the conservative quantity CO_2-acidity defined as

$$CO_2 - Acy = C_T - Alk \tag{11}$$

so that, by substitution:

$$\frac{d [CO_2 - Acy]}{dt} = K_a ([CO_2]_s - [CO_2]) + S_{CO_2 - Acy} \tag{12}$$

where S_{C_T} and $S_{CO_2 - Acy}$ are the external sources and sinks of total inorganic carbon and CO_2-acidity that may exist. K_a is the CO_2 rate constant for gas exchange and can be estimated from the O'Connor and Dobbins [5] reaeration formula. Wet and dry depositional loadings are included in the source term, S_{C_T} and $S_{CO_2 - Acy}$. For low pH, it is clear from the chemical equilibria that:

$$C_T = [CO_2] + [HCO_3^-] + [CO_3^=] \simeq [CO_2]$$

And since $CO_2-Acy = [CO_2] -[CO_3^=] - [OH^-] + [H^+]$ there exists a pH range for which $CO_2 \cdot Acy \simeq [CO_2]$. The accuracy of these approximations is within 10% and acceptable for most engineering purposes (Figure 6). For pH $\leqslant$ 4.3, the total inorganic carbon equation (1) can be written using $C_T \simeq CO_2$ with the result:

$$\frac{d [CO_2]}{dt} = K_a ([CO_2]_s - [CO_2]) + S_{C_T} \tag{13}$$

For $4.3 \leqslant$ pH $\leqslant 7.5$, the CO_2-acidity equation (12) can be written

$$\frac{d [CO_2]}{dt} = K_a ([CO_2]_s - [CO_2]) + S_{CO_2-Acy} \tag{14}$$

These equations form the basis for the analysis of a wide range of acidity problems in natural waters since they are linear in CO_2 and therefore are easily solved. Once the CO_2, C_T, and/or CO_2-Acy concentrations are known, the pH may be calculated from equilibrium considerations.

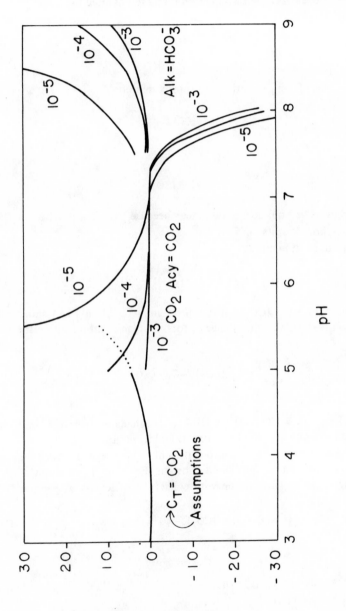

Figure 6. Percentage error vs pH.

For pH >7.5 the differential equation is still represented by:

$$\frac{dC_T}{dt} = K_a \left([CO_2]_s - [CO_2]\right) + S_{C_T} \tag{15}$$

However, in this pH range, neither the approximation:

$$C_T \approx [CO_2]$$

nor

$$[CO_2-Acy] \approx [CO_2]$$

is appropriate. The equation is nonlinear because CO_2 is a function of C_T, alkalinity, carbonate anion and pH.

Since it can be shown that:

$$[CO_2] = C_T - Alk + CO_3^= - H^+ + OH^- \tag{16}$$

we may substitute for $[CO_2]$ in the differential Equation 15. Assuming no external sources or sinks, the change of total inorganic carbon with respect to time is:

$$\frac{dC_T}{dt} = K_a \left([CO_2]_s - C_T - \phi(t)\right) \tag{17}$$

where $\phi(t) = -Alk + CO_3^= - H^+ + OH^-$. ϕ is calculated from the total acid loadings to the ecosystem by direct runoff and tributary inputs.

If a stream or river is divided into sufficiently short segments such that the alkalinity and pH can be considered constant, Equation 17 is simply a first-order, linear differential equation with constant coefficients. The solution is:

$$C_T = C_{T_0} \exp(-K_a t) + ([CO_2]_s - \phi)(1 - \exp(-K_a t)) \tag{18}$$

where

$$\phi = -Alk + CO_3^= - H^+ + OH^-$$
$$[CO_3^=] = K_a[Alk]/[H^+]$$

Once C_T is known, carbon dioxide and pH may be calculated from Equation 16.

In the expression for carbonate, the assumption is made that alkalinity is approximately equal to the bicarbonate ion. The assumption restricts the analysis to pH <9.0, but this pH range includes most values found in natural waters. The error in the assumption is given in Figure 6.

An excellent hypothetical example of increasing pH and degasification after an acid rainfall event is presented in Figure 7. Actual data of this sort were not available, but an example problem at nearly neutral pH was found.

Example Problem: Madison River

The diurnal pH at six sampling stations for the Madison River in Yellowstone Park was reported by Wright and Mills [7]. The data at 6:00 a.m. provided an example of stream data at pH <8.0. Field CO_2 concentrations were calculated from the pH and alkalinity data. To solve the differential equation (12) the assumption was made that CO_2-Acy $= CO_2$. The respiration measurements of Wright and Mills were included in the analysis and the solution to the CO_2 deficit equation was:

$$D = D_0 \exp(-K_a t) - \frac{P-R}{HK_a}[1 - \exp(-K_a t)] \quad (19)$$

where

D = CO_2 deficit
P = production sink of CO_2
R = respiration source of CO_2 (Table 2)
H = mean depth
t = travel time

Table 2. Madison River Segment Parameters

x (mi)	$\bar{U}$ (mi/day)	H (m)	Ka (day^{-1})	R (g/m^2-day)
1.56	30.6	0.59	6.65	1.0
3.64	19.1	1.09	2.13	0.4
6.60	29.6	0.44	10.2	0.0
9.66	36.2	0.36	15.1	0.1

The solution to Equation 19 using the O'Connor and Dobbins [5] reaeration formula to compute K_a is presented in Figure 8. Note the reasonable fit of the model to the field data.

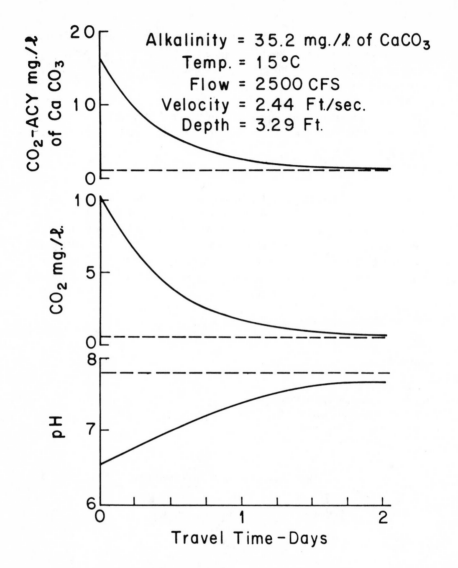

Figure 7. CO_2 exchange after acidification (from [6]).

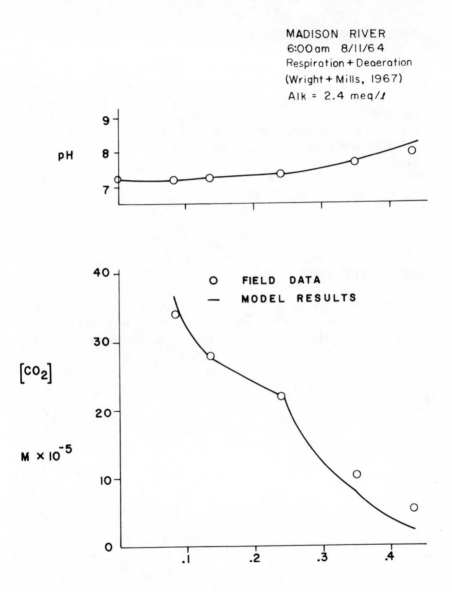

Figure 8. Madison River example.

CONCLUSIONS AND RECOMMENDATIONS

The trickle-down model proved to be a conceptual approach of significant value in a regional acid rainfall assessment. Much more effort is needed if it is to be used as a predictive tool. Specifically we plan to improve the weathering rates and kinetics and eventually to link the model into an atmospheric transport grid system. Metal dissolution and nutrient availability questions could be addressed within the constructs of this mass balance approach.

The simplifications introduced into the acid rainfall event model aided in solving a rather complex problem. With sufficient field data, it should be possible to calibrate and verify this model.

ACKNOWLEDGMENTS

We thank Professor Dominic M. DiToro for valuable input and collaboration on the Madison River example. Dr. Schnoor was supported as a University of Iowa Faculty Scholar during the tenure of this research.

REFERENCES

1. Glass, G. E., and O. L. Loucks. "Impacts of Airborne Pollutants on Wilderness Areas Along the Minnesota-Ontario Border," U.S. EPA Ecological Research Series (in press).
2. Oakes, E. L., and R. D. Cotter. "Water Resources of Wisconsin Upper Wisconsin River Basin," Atlas HA–536, U.S. Geological Survey, Reston, VA (1975), 3 pp.
3. Edzwald, J. K., and J. V. DePinto. "Recovery of Adirondack Acid Lakes with Fly Ash Treatment," Final Report to Engineering Foundation RC–A–76–4 (1978), 68 pp.
4. Kunelius, D. "Acid Rain, Is Time Running Out for Wisconsin Lakes?" *Wisconsin Sportsman* (July–August 1980), pp. 34–39.
5. O'Connor, D. J., and W. E. Dobbins. "Mechanism of Reaeration in Natural Streams," *Trans. Am. Cos. Civil Eng.* 123:655 (1958).
6. Shane, R. M. "Riverine Recreational Development-Mathematical Modeling," Department of Civil Engineering, Carnegie-Mellon University, Pittsburgh, PA (1974), pp. 24–40.
7. Wright, R., and T. Mills. "Diurnal Variations in pH of the Madison River, Yellowstone Park," *Limnol. Oceanog.* 12:367 (1967).

CHAPTER 14

ACIDIFICATION OF RAIN IN THE PRESENCE OF SO_2, H_2O_2, O_3 AND HNO_3

John H. Overton, Jr.

Environmental Sciences
Northrop Services, Inc.
Research Triangle Park, North Carolina 27709

Jack L. Durham

Environmental Sciences Research Laboratory
U.S. Environmental Protection Agency
Research Triangle Park, North Carolina 27711

INTRODUCTION

Odén [1] has performed an analysis of the data from the European wet deposition monitoring stations and observed the following annual trends for each station since the mid-1950s: (1) pH has declined, (2) sulfate has increased and (3) nitrate has increased. At present, there is not a comparable long-term data base for precipitation in the United States. Likens and Bormann [2] have reported observing from 1964 to 1972 an upward trend in annual wet deposition of nitrate and hydrogen ion and a downward trend for sulfate. They also noted similar trends of sulfate and nitrate at Geneva and Ithaca, New York.

The projected growth in the energy sector of the economy in the Ohio River Valley should be expected to increase acid precipitation over downwind areas, especially the northeastern United States. Relative to 1972 emissions in the valley, it is expected that by the year 2000 the emissions will change for SO_2 by a factor of -1 to 7% and for NO_x by a factor of 68 to 93% [3]. Thus, for the next 20 years the role of NO_x in acidifying rain may likely become more important.

The important mechanisms by which rain is acidified are not yet established, but they include oxidation of SO_2 and NO_x in cloud droplets, nucleation of droplets from acid particles, and scavenging of free gaseous acids (HNO_3 and HCl) by droplets. Holt et al. [4] used the technique of oxygen isotopy to investigate the pathway of SO_4^{2-} formation in raindrops; they concluded that oxidation in the droplet was always the predominant pathway. The principal oxidant is not known. However, using available reaction rate data, we have modeled the SO_4^{2-} formation rate for a subcloud event and found that ambient O_3 is much more effective than dissolved catalysts [5]. That work did not consider oxidation by H_2O_2. It is now known [6] that $H_2O_{2(g)}$ may be present in the northeastern United States atmosphere in concentrations as high as 40 ppb. Also, recent experimental evidence [7] on the reaction kinetics of $SO_{2(aq)}$ and $H_2O_{2(aq)}$ to yield H_2SO_4 overcomes Dasgupta's [8] concerns of the previous work of Penkett et al. [9].

At this time, there is no experimental evidence that indicates the predominant pathway for NO_3^- formation in raindrops. Durham et al. [10] recently modeled the effect of NO_2 (10 ppb) interacting with raindrops to yield NO, N_2O_3, N_2O_4 and HNO_2, followed by oxidation reactions to form HNO_3. That system was found to be unimportant. The only pathway yielding significant NO_3^- in raindrops was the absorption of free $HNO_{3(g)}$; free $HNO_{3(g)}$ is formed principally through the reaction of NO_2 and HO and often attains concentrations > 10 ppb in the northeastern United States [11].

In this chapter, we investigate: (1) the relative effectiveness of $H_2O_{2(aq)}$ and $O_{3(aq)}$ for oxidizing $SO_{2(aq)}$ to yield H^+ and SO_4^{2-}, and (2) the role of $HNO_{3(g)}$ in acidifying raindrops and influencing SO_4^{2-} formation. The important chemical behavior of rain in the presence of oxidants HNO_3 and SO_2 was determined by simulations using the physical model and chemical reaction module that are described in the following sections.

PHYSICAL MODEL

Our physical model of a rain event, raindrops and mass transfer are described elsewhere [5, 10]; thus, an abbreviated description is presented here.

The atmosphere has been divided into two regions, shown in Figure 1. Raindrops are formed in the upper region in the presence of CO_2 and other compounds which establish the initial pH. The drops enter and fall through a stable polluted region at a constant velocity. In the polluted region are the trace gases CO_2, O_3, SO_2, HNO_3 and H_2O_2. As the drop falls, gases are

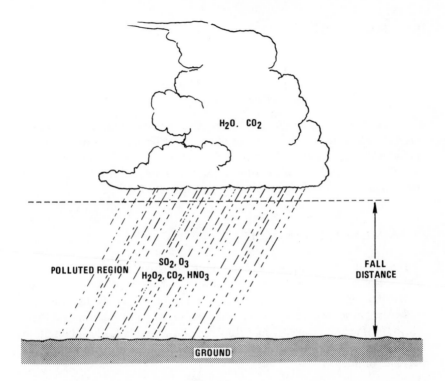

Figure 1. Environmental model. The raindrops form above and fall through the polluted zone. In the polluted zone, they absorb reactive gases which produce SO_4^{2-} and NO_3^-.

absorbed, react and produce SO_2^{2-}, NO_3^- and other species. A raindrop is taken to be a uniform sphere composed of water and trace quantities of O_3, SO_2, CO_2, HNO_3, H_2O_2 and their products. Concentrations are assumed to be uniformly distributed at all times throughout the drop, (i.e., no concentration gradients in the drop). There is no coagulation or fragmentation of drops. Raindrop temperature is assumed to be in equilibrium with an isothermal atmosphere at $25°C$ (a choice dictated by the available chemical rate constant data). The raindrop size distribution is assumed to be independent of height and time and to be distributed according to the distribution of Best [12] truncated at a lower radius limit of 0.05 mm.

The rate, per unit fall distance, at which a gaseous species (e.g., O_3, HNO_3,

H_2O_2, SO_2 and CO_2) crosses the gas-water interface of a drop of radius R is given by

$$\left.\frac{d[s]_L}{dz}\right|_{transfer} = \frac{30.0k_g}{Ru}([s]_g - H_s[s]_L)$$

The notation is presented in a section at the end of this chapter. The species, s, is assumed to be distributed uniformly through the drop. The mass transfer coefficient, k_g, is obtained from the Frössling correlation [13]. Values of u, the fall velocity, are obtained from a formula by Markowitz [14]. Henry's law constants, H_s, for the gaseous species are listed in Table 1.

CHEMICAL REACTION MODULE

The kinetic reactions used for this work are given in Table 2. Equations 1-6 are the reversible reactions for the CO_2-SO_2-HNO_3-H_2O system. The irreversible oxidation steps are given in Equations 7 to 10. Reaction 10, H_2O_2 oxidation, is a compromise between the works of Martin and Damschen [7], Penkett et al. [9] and Hoffman and Edwards [19] and results in rates no more than twice or no less than one half the reported values in the pH range of interest.

MATHEMATICAL FORMULATION AND INTERPRETATION

The concentration (as a function of fall distance) of each chemical species within a drop of a given radius was obtained by numerically integrating the set of coupled nonlinear differential equations derived from the kinetic and mass transport equations. This procedure was repeated for several different radii, and with the application of the Best [12] raindrop dize distribution,

Table 1. Henry's Law Constants

Species	H_s	Source of H_s Values
CO_2	1.2	Perry and Chilton [15]
O_3	3.36	Perry and Chilton [15]
SO_2	0.0332	Hales and Sutter [16]
HNO_3	0.46E-6	Abel and Nuesser [17] and McKay [18]
H_2O_2	0.557E-6	Martin and Damschen [7]

Table 2. Kinetic Mechanism of CO_2-SO_2-HNO_3-O_3-H_2O_2-Aqueous Phase System Used for Raindrops[a]

1. (H_2O)	$\xrightarrow{1.3E-3}$ $\xleftarrow{1.3E+11}$	$H^+ + OH^-$	Eigen et al. [20]
2. $CO_2 + (H_2O)$	$\xrightarrow{2.4E-2}$ $\xleftarrow{4.92E+4}$	$HCO_3^- + H^+$	Himmelblau and Babb [21] (by extrapolation to $25^\circ C$)
3. HCO_3^-	$\xrightarrow{1.0E-4}$ $\xleftarrow{1.4E+4}$	$CO_2 + OH^-$	Eigen et al. [20]
4. $SO_2 + (H_2O)$	$\xrightarrow{3.4E+6}$ $\xleftarrow{2.0E+8}$	$H^+ + HSO_3^- (20^\circ C)$	Eigen et al. [20]
5. HSO_3^-	$\xrightarrow{1.0E+4}$ $\xleftarrow{1.0E+11}$	$SO_3^{2-} + H^+$	Erickson and Yates [22] Erickson et al. [23]
6. HNO_3	$\xrightarrow{2.2E+9}$ $\xleftarrow{1.0E+8}$	$H^+ + NO_3^-$	McKay [18]
7. $O_3 + SO_2$	$\xrightarrow{5.9E+2}$	$2H^+ + SO_4^{2-}$	Erickson et al. [23]
8. $O_3 + SO_3^{2-}$	$\xrightarrow{2.2E+9}$	SO_4^{2-}	Erickson et al. [23]
9. $O_3 + HSO_3^-$	$\xrightarrow{3.1E+5}$	$H^+ + SO_4^{2-}$	Erickson et al. [23]
10. $H_2O_2 + HSO_3^- + H^+$	$\xrightarrow{7.3E+7}$	$2H^+ + SO_4^{2-}$	Hoffman and Edwards [19] Penkett et al. [9] Martin and Damschen [7]

[a]Units are in liters, moles, seconds.

simulated "samples" of rainwater were computed. (A more detailed account of the mathematical formulation is found in Overton et al. [5] and Durham et al. [10].)

The "samples" of the simulated rainfall are: (1) "collected" sequentially, each one for several minutes, (2) isolated from ambient factors so there is no mass transfer with the atmosphere, (3) allowed to continue reacting. In addition, the SO_4^{2-} and pH values reported are those that would result if *all* the S(IV) were transformed to SO_4^{2-} (by, for example, a slow catalytic process over a long period of time due to storage in the collector or sample vials).

RESULTS AND DISCUSSIONS

The relative effectiveness of $H_2O_{2(aq)}$ and $O_{3(aq)}$ for oxidizing $SO_{2(aq)}$ and the role of HNO_3 are demonstrated by the results of simulations of 1000-m, 10-mm/hr sequentially sampled rain events. These results have been divided into five cases. The first three cases are for rain events of 5-min duration in which the air pollutant concentrations are assumed to remain constant as raindrops with an initial pH = 5.6 fall through the polluted zone. The fourth case is for an extended rain event in which gas pollutant concentrations are lowered during the event in response to washout by the rain. The

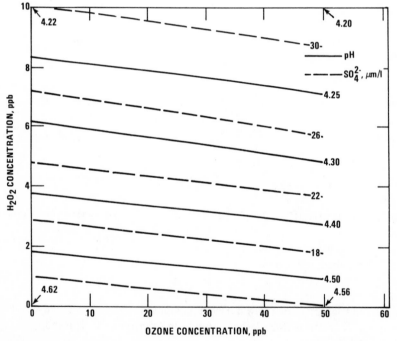

Figure 2. The effect of ambient H_2O_2 and O_3 on pH and sulfate in rain. Plotted are isopleths of pH (——) and SO_4^{2-} (--- micromoles/liter) for ambient values of H_2O_2 and O_3. The data are from a model simulation of a subcloud scavenging rain event. Conditions: fall distance = 1000 m, initial pH = 5.63, ambient $[SO_2]$ = 10 ppb, ambient $[HNO_3]$ = 0 ppb.

fifth case compares the effect of preacidification (pH = 4) with the events for which the initial pH - 5.6.

Case I. Relative Effectiveness of H_2O_2 and O_3 on Acidification and SO_4^{2-} Formation. For a 5-min period in a steady rain, we calculated the pH and $[SO_4^{2-}]$ of drops with an initial pH = 5.6 that fall through a polluted zone (1000 m). The results are shown in Figure 2 for which $HNO_{3(g)}$ is absent, $[SO_{2(g)}]$ = 10 ppb, and H_2O_2 (0-10 ppb) and O_3 (0-50 ppb) are variable. As can be seen in the slight negative slopes of the isolines, the relative effectiveness of H_2O_2 to O_3 is about 12 (ppb/ppb) for lowering pH and for forming SO_4^{2-} for the conditions listed in Figure 2. For example, note the difference for $[H_2O_{2(g)}]$ = 10 ppb and $[O_{3(g)}]$ = 0 and 50 ppb: in the absence of O_3 the pH is 4.22, but when O_3 is increased to 50 ppb, the pH becomes 4.20. That is, 10 ppb of H_2O_2 drops the pH by 0.4 from 4.62 (at $[H_2O_2]$ = $[O_3]$ = 0) to 4.22, but an additional 50 ppb O_3 results only in a drop in pH by 0.02 from 4.22 to 4.20.

In the presence of $HNO_{3(g)}$ (not shown here), the relative effectiveness of H_2O_2 to O_3 increases. That is because H^+ inhibits the O_3 oxidation of dissolved SO_2 species, but not the H_2O_2 oxidation. In the absence and in the presence of $HNO_{3(g)}$, H_2O_2 is a more effective oxidizer than O_3.

Case II. Relative Effectiveness of HNO_3 and SO_2 on Acidification and SO_4^{2-} Formation. Two subcases, $[H_2O_{2(g)}]$ = 0 and 10 ppb, are considered in order to compare the relative effectiveness of HNO_3 and SO_2 in the presence of $[O_{3(g)}]$ = 50 ppb. The calculated pH, $[SO_4^{2-}]$, and $[NO_3^-]/[SO_4^{2-}]$ isopleths for a 5-min rain sample are shown in Figure 3 for $[H_2O_2]$ = 0. This system has been discussed in detail elsewhere [10]. The small (<-1.) negative slope of the pH isolines demonstrates the dominance of $HNO_{3(g)}$ in controlling the acidity of this system. Also, $HNO_{3(g)}$ is effective in suppressing $[SO_4^{2-}]$ formation due to O_3 oxidation. For example, at $[SO_{2(g)}]$ = 15 ppb, note that as $[HNO_{3(g)}]$ ranges from 0 to 10 ppb, $[SO_4^{2-}]$ falls from 17.5 μM to < 12.5 μM, and pH declines by ~ 0.3.

The isolines for this system change dramatically when $H_2O_{2(g)}$ is present at 10 ppb, as is shown in Figure 4. The pH isolines become steeper, indicating stronger dependence on $[SO_{2(g)}]$, and are significantly displaced to lower values. The $[SO_4^{2-}]$ isolines are essentially vertical, indicating that the SO_4^{2-} formation due to H_2O_2 oxidation is strongly dependent on $[SO_{2(g)}]$ and only weakly dependent on $HNO_{3(g)}$. For HNO_3 to contribute over 50% to the acidification ($[NO_3^-]/[SO_4^{2-}] \geqslant 2$), then the $HNO_{3(g)}$ concentration must exceed that of SO_2 by a factor of ~ 1.3. For $[SO_{2(g)}] \leqslant [H_2O_2]$ = 10 ppb, the effectiveness of SO_2 is ~ 1.3 of that of $HNO_{3(g)}$ for acidifying rain; however, as $[SO_{2(g)}]$ > $[H_2O_{2(g)}]$, the relative effectiveness of $[SO_{2(g)}]$

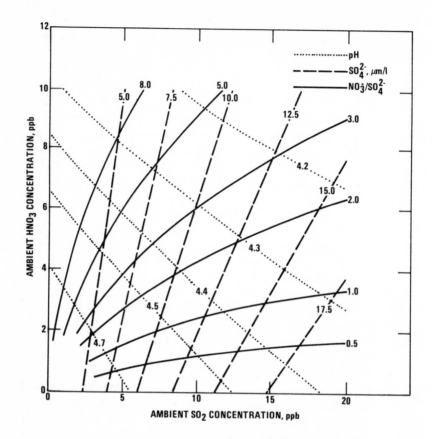

Figure 3. The effect of ambient SO_2 and HNO_3 on pH, sulfate and nitrate in rain. Plotted are isopleths of pH (••••••••), SO_4^{2-} (- - - μmol/l), and the ratio (NO_3^-/SO_4^{2-}) (———) for ambient values of SO_2 and HNO_3. The data are from a model simulation of a subcloud scavenging rain event. Conditions: fall distance = 1000 m, initial pH = 5.63, ambient $[O_3]$ = 50 ppb, ambient $[H_2O_2]$ = 0 ppb. The absorption of HNO_3 by the falling droplets lowers the pH and retards the formation of SO_4^{2-} due to liquid phase reaction between dissolved O_3 and SO_2 species.

declines. If $[SO_{2(g)}] > 7.5$ ppb, then $[HNO_{3(g)}]$ must be > 10 ppb in order for $HNO_{3(g)}$ to dominate the acidification.

The behavior of the systems considered in Case II demonstrates that for $[O_{3(g)}] = 50$ ppb: (a) $HNO_{3(g)}$ in the absence of H_2O_2 is more effective than SO_2 in acidifying rain and may limit SO_4^{2-} formation, but (b) in the presence of $H_2O_{2(g)}$, $SO_{2(g)}$ is more effective and the SO_4^{2-} formation rate is weakly dependent on $HNO_{3(g)}$.

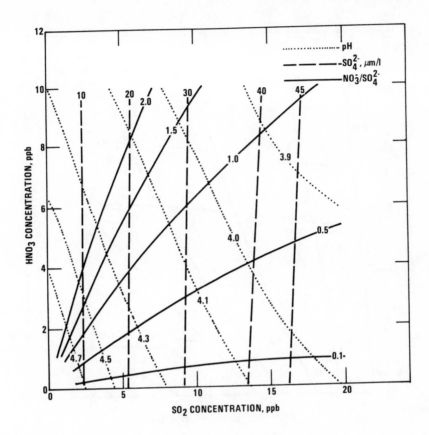

Figure 4. The effect of ambient SO_2 and HNO_3 on pH, sulfate and nitrate in the presence of H_2O_2. Plotted are isopleths of pH ($\cdots\cdots$), SO_4^{2-} ($---$ micromoles/liter), and the ratio (NO_3^-/SO_4^{2-}) ($\longrightarrow$) for ambient values of SO_2 and HNO_3. The data are from a model simulation of a subcloud scavenging rain event. Conditions: fall distance = 1000 m, initial pH = 5.63, ambient $[O_3]$ = 50 ppb, ambient $[H_2O_2]$ = 10 ppb. The absorption of HNO_3 by the falling droplets lowers the pH and slightly retards the formation of SO_4^{2-}.

Case III. Relative Effectiveness of H_2O_2 and HNO_3 on Acidification and SO_4^{2-}-Formation. Two subcases are shown here in order to compare the relative effectiveness of H_2O_2 and HNO_3 (for $[O_{3(g)}]$ = 50 ppb) in the presence of $[SO_{2(g)}]$ = 10 ppb (see Figure 5) and $[SO_{2(g)}]$ = 20 ppb (see Figure 6). This case is analogous to Case II. The influence of reaction 10 is being demonstrated. In Case II, $[HSO_3^-]$ from reaction 10 was the variable and in Case III, it is $[H_2O_2]$. In Figure 5, $[H_2O_{2(g)}] \leqslant [SO_{2(g)}]$ = 10 ppb, and H_2O_2 and HNO_3 have approximately equal effectiveness for acidification; note however

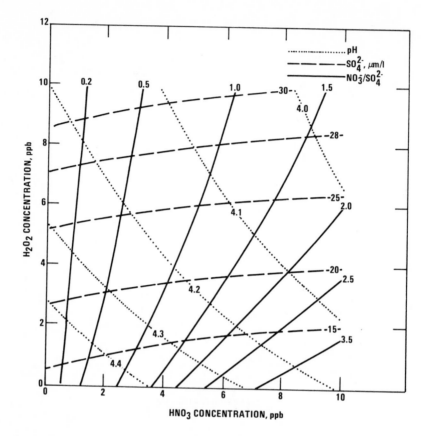

Figure 5. The effect of ambient H_2O_2 and HNO_3 on pH, sulfate and nitrate in rain. Plotted are isopleths of pH ($\circ\bullet\bullet\bullet\bullet\circ$), SO_4^{2-} ($----$ micromoles/liter), and the ratio (NO_3^-/SO_4^{2-}) ($——$). The data are from a model simulation of a subcloud scavenging rain event. Conditions: fall distance = 1000 m, initial pH = 5.63 , ambient $[O_3]$ = 50 ppb, ambient $[SO_2]$ = 10 ppb.

that SO_4^{2-} formation is only weakly dependent on HNO_3 (as observed for Case II). In Figure 6, $[H_2O_2{_{(g)}}] \leqslant 10$ ppb and $[SO_2{_{(g)}}] = 20$ ppb; due to this increase in $[SO_2{_{(g)}}]/[H_2O_2{_{(g)}}]$, the $SO_2{_{(aq)}}$ oxidation by $H_2O_2{_{(aq)}}$ become more important and HNO_3 less important to the acidification. That is illustrated by the increase (less negative) in slope of the pH isolines and their shift to lower values as $[SO_2{_{(g)}}]$ is increased from 10-20 ppb.

This case is analogous to Case II, and $H_2O_2{_{(g)}}$ is more effective than $HNO_3{_{(g)}}$ in acidifying rain.

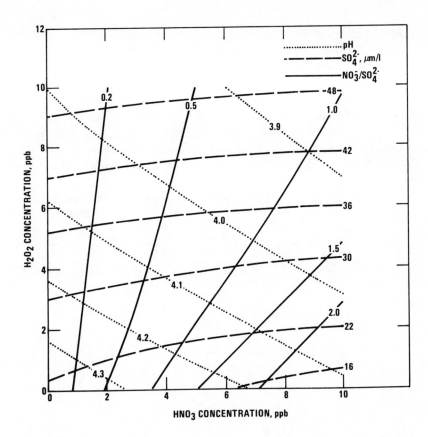

Figure 6. The effect of ambient H_2O_2 and HNO_3 on pH, sulfate and nitrate in rain. Plotted are isopleths of pH (•••••••), SO_4^{2-} (– – – micromoles/liter), and the ratio (NO_3^-/SO_4^{2-}) (————). The data are from a model simulation of a subcloud scavenging rain event. Conditions: fall distance = 1000 m, initial pH = 5.63, ambient $[O_3]$ = 50 ppb, ambient $[SO_2]$ = 20 ppb.

Case IV. Extended Rain Event. Figures 7 and 8 illustrate the interplay between rain concentrations and ambient gas concentrations for simulated 10-mm/hr, 1000-m, 60-min rain events. In Figure 7, $[H_2O_2(g)] = 0$, we note the rapid decrease of $[HNO_3(g)]$ accompanied by an even more rapid decrease of rainwater NO_3^-. On the other hand we observe a slight increase in SO_4^{2-} due to an increase in pH (gaseous O_3 and SO_2 decrease at a much slower rate than HNO_3). Figure 8, initial $[H_2O_2(g)] = 5$ ppb, shows the nitrogen species behaving the same as in the absence of H_2O_2 along with the expected decrease of H_2O_2 due to its removal by the rain. Again the pH increases; however, in contrast to the $[H_2O_2(g)] = 0$ case, SO_4^{2-} is initially twice as much and *decreases* (42% in 60 minutes) as the event progresses. This decrease is obviously due to the removal of H_2O_2 from the atmosphere; note that the SO_4^{2-} curve "follows" the ambient H_2O_2 curve.

As pointed out, the behavior of HNO_3 and NO_3^- is the same in both cases (Figures 7 and 8). For the kinetic mechanism used $[NO_3^-]$ depends only on the fall distance and $[HNO_{3(g)}]$. This is due to the very low Henry's law constant (that results in negligible HNO_3 back pressure) and to the dissociation of HNO_3 (Equation 6, Table 2) in such a way that most of the nitrogen is in the form of NO_3^- and is thus insensitive to pH for a given quantity of nitrogen compounds.

The ratio, $[NO_3^-]/[SO_4^{2-}]$, (not shown) decreases in both cases; however,

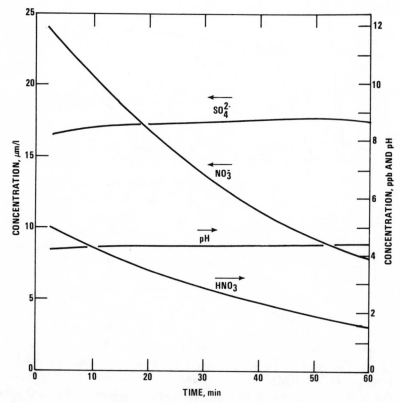

Figure 7. Smoothed variations of SO_4^{2-}, NO_3^- and pH in sequentially collected 5-minute samples of rainwater and of the ambient gas HNO_3 as a function of time during a 1000-m, 10-mm/hr, 60-minute simulated subcloud scavenging rain event. Initial conditions: pH = 5.63, $[O_{3(g)}] = 50$ ppb, $[SO_{2(g)}] = 20$ ppb, $[HNO_{3(g)}] = 5$ ppb, $[H_2O_{2(g)}] = 0$ ppb.

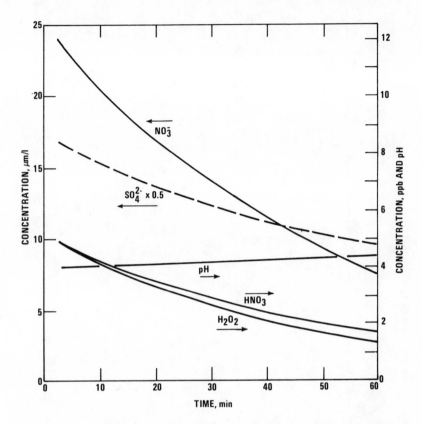

Figure 8. Smoothed variations of SO_4^{2-}, NO_3^- and pH in sequentially collected 5-min samples of rainwater and of the ambient gases HNO_3 and H_2O_2 as a function of time during a 1000-m, 10-mm/hr, 60-min simulated subcloud scavenging rain event. Initial conditions: pH = 5.63, $[O_{3(g)}]$ = 50 ppb, $[SO_{2(g)}]$ = 20 ppb, $[H_2O_{2(g)}]$ = $[HNO_{3(g)}]$ = 5 ppb.

due to the near constancy of SO_4^{2-} in the $[H_2O_2]$ = 0 case, the ratio decreases more rapidly than in the case when H_2O_2 is present.

Case V. Effect of Preacidification. Raindrops may be acidified prior to entering the polluted zone. Here we assume that occurs and that the raindrops have an initial pH = 4 and $[SO_4^{2-}]$ = 50 μM prior to entering the polluted zone. The effect on pH and SO_4^{2-} formation for select values of HNO_3, O_3 and H_2O_2 in the presence of $[SO_{2(g)}]$ = 10 ppb is illustrated in Table 3 where values of the *change* in pH and SO_4^{2-} with respect to their initial values are shown. From the table we note the following: (1) preacidification limits further changes in pH and (2) in the presence of H_2O_2, SO_4^{2-} production is

Table 3. Effect of Preacidification (SO$_2$ = 10 ppb)[a]

Ambient Concentration (ppb)			ΔpH		ΔSO$_4^{2-}$, μM	
HNO$_3$	O$_3$	H$_2$O$_2$	Initial pH: 5.6	4.0	5.6	4.0
0	0	0	-1.0	0	12.0	2.0
0	50	0	-1.1	-0.05	14.0	6.6
10	0	0	-1.4	-0.2	6.4	1.6
10	50	0	-1.4	-0.2	8.6	1.9
0	0	10	-1.4	-0.2	30.0	26.0
0	50	10	-1.4	-0.2	32.0	27.0
10	0	10	-1.6	-0.3	29.0	26.0
10	50	10	-1.7	-0.3	30.0	27.0

[a]Fall height = 1000 m, rain rate = 10 mm/hr.

independent of preacidification (preacidification does limit SO_4^{2-} production by O_3 oxidation). However we point out that, as the H_2O_2 oxidation rate is independent of pH, conclusion (1) would no doubt have to be modified somewhat for fall heights of > 1000 m; since, unlike the O_3 oxidation step, H_2O_2 will continue to oxidize S(IV). A fact to note is that a change of pH from 4.6 to 3.6 requires ten times as much hydrogen ions as the change from 5.6 to 4.6 and thus roughly a fall distance 10 times as much as that for a 5.6 to 4.6 pH change. Since the mass transfer rate is limited the quantity of mass (SO_2, O_3, H_2O_2, etc.) that can be absorbed to produce $[H^+]$ is also limited; thus, the *effective* results of preacidification with respect to pH has the appearance of quenching the reaction.

CONCLUSIONS

As a result of this investigation of the effect of SO_2, H_2O_2, O_3 and HNO_3 on the production of acidity in rain we conclude that:

1. In the absence and in the presence of HNO_3, H_2O_2 is a more effective oxidizer than O_3.
2. HNO_3, in the absence of H_2O_2, is more effective than SO_2 in acidifying rain and may limit SO_4^{2-} formation.
3. In the presence of H_2O_2, SO_2 may be more effective than HNO_3 in acidifying rain; in this case the SO_4^{2-} formation rate is weakly dependent on HNO_3.
4. NO_3^- formation, for the chemical mechanism used, depends mainly on ambient HNO_3, fall distance and rain rate, but not on pH.
5. Acidification prior to entering polluted zone inhibits the further production of acidity (based on pH) and SO_4^{2-} in the absence of H_2O_2.

Meaningful comparisons of our simulations to field data can not be made as the model is limited in its assumptions and in its description of rain events [5, 10]. However, the chemistry module used in these simulations contains many of the important reactions expected to take place in rainwater, and as such the simulations do increase our understanding of the important chemical reactions in acidifying rain.

NOTATION

H   Henry's law constant, dimensionless. Equilibrium ratio of gas phase concentration to liquid phase concentration of same species.

R   drop radius, mm

aq   aqueous

g   indicates gas phase

hr   hour

k_g gas-phase mass transfer coefficient, cm/sec

l   liter

m   meter

mm   millimeter

s   indicates a chemical species such as SO_2, HNO_3, H^+, HSO_3^-, etc.

u   fall velocity, m/sec

z   fall distance, m

ℓ indicates liquid phase

[]   molar concentration

μM 10^{-6} mol/l

REFERENCES

1. Odén, S. "The Acidity Problem—An Outline of Concepts," in *Proceedings of the First International Symposium on Acid Precipitation and the Forest Ecology, May 12-15, 1975, Columbus, OH.* USDA Forest Service General Technical Report NE-23, 1-30.

2. Likens, E. and F. H. Bormann. "Acid Rain: A Serious Regional Environmental Problem," *Science* 184:1176-1179 (1974).

3. Stukel, J. J. and B. R. Keenan. "Ohio River Basin Energy Study. ORBES Phase I: Interim Findings," U.S. EPA Report EPA-600/7-77-120 (1977), pp. 46-47.

4. Holt, B. D., P. T. Cunningham and R. Kumar. "Oxygen Isotopy of Atmospheric Sulfates," *Environ. Sci. Technol.* (in press).

5. Overton, J. H., V. P. Aneja and J. L. Durham. "Producton of Sulfate in Rain and Raindrops in Polluted Atmospheres," *Atmos. Environ.* 13: 355-367 (1979).

6. Gay, B. W. and J. J. Bufalini. "Hydrogen Peroxide in the Urban Atmosphere," in *Photochemical Smog and Ozone Reactions,* ACS Advances in Chemistry Series, Vol. 113, (1972), pp. 255-263.

7. Martin, L. R. and D. E. Damschen. "Aqueous Oxidation of Sulfur Dioxide by Hydrogen Peroxide at Low pH," *Atmos. Environ.* accepted for publication.

8. Dasgupta, P. K. "The Importance of Atmospheric Ozone and Hydrogen Peroxide in Oxidizing Sulfur Dioxide in Cloud and Rainwater—Further Discussion (Penkett et al. 1979)," *Atmos. Environ.* 14:620-621 (1980).

9. Penkett, S. A., et al. "The Importance of Atmospheric Ozone and Hydrogen Peroxide in Oxidizing Sulphur Dioxide in Cloud and Rainwater," *Atmos. Environ.* 13:123-127 (1979).

10. Durham, J. L., J. H. Overton, Jr. and V. P. Aneja. "Influence of Gaseous Nitric Acid on Sulfate Production and Acidity in Rain," *Atmos. Environ.* (in press).

11. Spicer, C. W. "Measurement of Gaseous HNO_3 by Electrochemistry and Chemiluminescence," in *Current Methods to Measure Atmospheric Nitric Acid and Nitric Acid Artifacts,* R. K. Stevens, Ed., U.S. EPA-

600/2-79-051 (Research Triangle Park, NC: U.S. Environmental Protection Agency, 1979), pp. 27–35.

12. Best, A. C. "The Size Distribution of Raindrops," *Quart. J. Roy. Meteorol. Soc.* 76:16–36 (1950).

13. Frössling, N. "The Evaporation of Falling Drops," *Gerlands Beits. Geophys.*, 52:170–216 (1938).

14. Markowitz, A. H. "Raindrop Size Distribution Expressions," *J. Appl. Meteorol.* 15:1029–1031 (1976).

15. Perry, R. H. and C. H. Chilton, Eds. *Chemical Engineer's Handbook.* 5th ed. (New York: McGraw-Hill Book Company, 1973).

16. Hales, J. M. and S. L. Sutter. "Solubility of Sulfur Dioxide in Water at Low Concentrations," *Atmos. Environ.* 1:997–1001 (1973).

17. Abel, E. and E. Neusser. "Über den Dampfdruck der Salptrigen Säure," *Monatshafte fuer Chemie,* 54:855–873 (1929).

18. McKay, H. A. C. "The Activity Coefficient of Nitric Acid, A Partially Ionized 1:1-Electrolyte," *Trans. Faraday Soc.* 52:1568–1573 (1956).

19. Hoffmann, M. R. and J. O. Edwards. "Kinetics of the Oxidation of Sulfite by Hydrogen Peroxide in Acidic Solution," *J. Phy. Chem.,* 79(20): 2096–2098 (1975).

20. Eigen, M., et al. "Rate Constants of Protolytic Reactions in Aqueous Solution," *Progress in Reaction Kinetics, Vol. 2,* G. Porter, Ed. (New York: MacMillan, 1964) pp. 287–318.

21. Himmelblau, D. M. and A. L. Babb. "Reaction Rate Constants by Radioactive Tracer Techniques," *Ind. Eng. Chem.* 51(11):1403–1408 (1959).

22. Erickson, R. E. and L. M. Yates. "Reaction Kinetics of Ozone with Sulfur Compounds," U.S. EPA–600/3-76-089 (1976), 69 pp.

23. Erickson, R. E., et al. "The Reaction of Sulfur Dioxide with Ozone in Water and its Possible Atmospheric Significance," *Atmos. Environ.* 11: 813–817 (1977).

CHAPTER 15

THE INFLUENCE OF GAS SCAVENGING OF TRACE GASES ON PRECIPITATION ACIDITY

Mahmoud Reda, Yusuf Adewuyi and Gregory R. Carmichael

Chemical and Materials Engineering Program
University of Iowa
Iowa City, Iowa 52242

INTRODUCTION

Rain is naturally slightly acidic (pH 5.6) as a result of absorption of carbon dioxide from the atmosphere. However, rains as acidic as pH 4 are commonly reported [1] and in the highly industrialized regions, rain has been analyzed to contain carbonic acid (H_2CO_3), nitric acid (HNO_3), hydrochloric acid (HCl) and sulfuric acid (H_2SO_4). These in turn are the legacies of ever-greater auto emissions, use of nitrogen fertilizer, and fossil fuel consumption by power plants, smelters, refineries and other chemical industries.

Combustion of fossil fuels and the conversion of fertilizer nitrogen by microorganisms in soil releases sulfur and nitrogen oxides into the atmosphere. These gases are the precursors of the aforementioned strong acids. On the other hand, the availability in the atmosphere of ammonia (NH_3), which originates largely from the decay of organic matter, appears to determine the extent to which these strong acids can be chemically neutralized.

Raindrops absorb atmospheric gases and thus provide an important mechanism for cleaning the atmosphere. This process is called precipitation scavenging. Precipitation scavenging is classified into rainout (processes within clouds) and washout (processes below clouds). Although precipitation scavenging constitutes an important sink for otherwise-noxious gases, the consequences of such a phenomenon may have far-reaching environmental impacts. Damage to agricultural crops, fish and other aquatic ecosystems, release of toxic metals from soil and subsequent contamination of drinking water, and deterioration of statuary and other man-made materials are only a few of the potential hazards of these acid rains.

With the anticipated increase in coal burning during the next few decades and in view of the possible hazards mentioned above, the importance of understanding in detail the varied aspects of precipitation scavenging cannot be overemphasized. Recently there have been a number of studies, both experimental and theoretical, dealing with the simultaneous absorption of SO_2 and other trace gases, and models that predict the composition of rain as a function of fall distance have been developed.

In Terraglio and Manganelli's [2] experimental study, distilled water samples in the form of surface films with only slight depth were exposed to SO_2 in the concentration range of 0.31–3.34 ppm. They found that the pH of the solution decreased from 6.1 to an equilibrium value of 3.6–4.1. They also found the rate of SO_2 absorption to be a function of SO_2 concentration and that 98.5% of the sulfite in solution was in the form of the bisulfite ion (HSO_3^-).

Hill and Adamowicz [3] developed a physicochemical model for SO_2 washout in which the mass transfer of SO_2 into well-mixed drops of rain with either strong acidic or basic background was considered. However, effects of NH_3, CO_2 and other trace gases were ignored and the bisulfite ion was assumed to be the predominant sulfur-bearing species in solution. Adamowicz [4] extended this model to include the simultaneous absorption of NH_3, CO_2 and catalyzed oxidation of the dissolved sulfur. For all drop sizes, NH_3 was found to increase the equilibrium pH and hence the rain's capacity for washout.

Results similar to those of Adamowicz were obtained by Overton et al. [5] in their model in which mass transfer of SO_2, O_3, NH_3 and CO_2 into ideal drops with initial pH of 5.56 and containing Fe(III) catalyst were considered. In addition to the above results, they found equilibrium pH of the drops to range from 4.2 to 6.6, depending on the drop size, fall distance and ambient concentration. In a similar work, Aneja et al. [6] found that the pH of rain varies from 3.9 to 6.4 and is independent of the initial pH. They found that large drops with initial pH of 10 remain basic and that the addition of NH_3 enhances sulfate production.

In all the above models, certain assumptions have been used. These include zero interfacial concentration or complete gas-phase-controlled mass transfer, linearized solubility and the predominance of a specific sulfur-bearing species. Carmichael and Peters [7] developed a model for treating simultaneous absorption and chemical reaction of SO_2 into aqueous solutions which eliminates most of these simplifying assumptions. The model predicts absorption rates and pH and utilizes the fact that the absorption can be described in terms of the physical mass transfer coefficient when the driving force is the total liquid phase S(IV) concentration. However, the effects of other trace gases were ignored.

In this chapter, results from new gas scavenging models are presented and discussed. First, the results obtained by extending the model of Carmichael and Peters to include the simultaneous absorption of NH_3, NO_x, HNO_3, CO_2 and HCl are presented. Effects of drop size, pollutant concentration, initial pH, fall distance, travel time and temperature are investigated. These results, as well as most other generalized treatments of washout, treat only the isothermal, no-growth droplet cases. However, nonisothermal effects are present to some extent in all gas absorption processes. Results discussed in the second part of the chapter demonstrate that these effects can be significant in washout calculations.

ABSORPTION OF SO_2, HNO_3, CO_2, NH_3 AND HCL BY WATER DROPLETS

In this section simultaneous trace gas absorption by raindrops is analyzed by extending the model of Carmichael and Peters [7]. The raindrops are assumed to be spherical with a constant temperature and fall vertically from cloudbase at their terminal velocities. At cloudbase the drops are assumed to be pollutant-free with a pH of 5.56 and the ambient concentrations of the trace gases are assumed to be spatially and temporally uniform. Mixing within the raindrops (due to internal circulation) is assumed rapid enough to provide a uniform concentration within the bulk of the droplet and all ionic equilibria resulting from the dissociation of absorbed gases are instantaneously established.

Solution Chemistry

Many absorbed gases dissociate in solution. This dissociation enables an increased amount of gas to enter the droplet by enhancing the concentration gradient driving force. The following reactions are considered.

$$H_2O \rightleftarrows H^+ + OH^-, \quad K_w = [H^+] [OH^-] \tag{1}$$

$$SO_2(g) + H_2O \rightleftarrows SO_2 \cdot H_2O, \quad [SO_2(g)] = H1 [SO_2 \cdot H_2O] \tag{2}$$

$$SO_2 \cdot H_2O \rightleftarrows H^+ + HSO_3^-, \quad K1 = [H^+] [HSO_3^-]/[SO_2 \cdot H_2O] \tag{3}$$

$$HSO_3^- \rightleftarrows H^+ + SO_3^{2-}, \quad K2 = [H^+] [SO_3^{2-}]/[HSO_3^-] \tag{4}$$

$$CO_2(g) + H_2O \rightleftarrows CO_2 \cdot H_2O, \quad [CO_2(g)] = H2 [CO_2 \cdot H_2O] \tag{5}$$

$$CO_2 \cdot H_2O \rightleftarrows H^+ + HCO_3^-, \quad K3 = [H^+] [HCO_3^-]/[CO_2 \cdot H_2O] \tag{6}$$

$$HCO_3^- \rightleftarrows H^+ + CO_3^{2-}, \quad K4 = [H^+] [CO_3^{2-}]/[HCO_3^-] \tag{7}$$

$$NH_3(g) + H_2O \rightleftarrows NH_3 \cdot H_2O, \quad [NH_3(g)] = H3 [NH_3 \cdot H_2O] \tag{8}$$

$$NH_3 \cdot H_2O \rightleftarrows NH_4^+ + OH^-, \quad K5 = [NH_4^+] [OH^-]/[NH_3 \cdot H_2O] \tag{9}$$

$$HNO_3(g) + H_2O \rightleftarrows HNO_3 \cdot H_2O, \quad [HNO_3(g)] = H4[HNO_3 \cdot H_2O] \tag{10}$$

$$HNO_3 \cdot H_2O \rightleftarrows H^+ + NO_3^- + H_2O, \quad K6 = [H^+] [NO_3^-]/[HNO_3 \cdot H_2O] \tag{11}$$

$$HCl(g) + H_2O \rightleftarrows HCl \cdot H_2O, \quad [HCl(g)] = H5[HCl \cdot H_2O] \tag{12}$$

$$HCl \cdot H_2O \rightleftarrows H^+ + Cl^- + H_2O. \quad K7 = [H^+] [Cl^-]/[HCl \cdot H_2O] \tag{13}$$

(Values for the dissociation constants and the Henry's law constants are listed in Table 1.)

The liquid solution also must satisfy the electrical-neutrality constraint, which for the overall system is

$$[NH_4^+] + [H^+]_T = [OH^-]_T + [HSO_3^-] + 2[SO_3^{2-}] + [HCO_3^-] + 2[CO_3^{2-}]$$
$$+ [NO_3^-] + [Cl^-] \tag{14}$$

Table 1. Equilibrium Constants (Molar) and Henry's Law Constants (mol/L gas / mol/L aqueous) at 25°C

$K_1 =$	0.017	$H1 = 3.3 \times 10^{-2}$
$K_2 =$	6.24×10^{-8}	$H2 = 1.2$
$K_3 =$	4.45×10^{-7}	$H3 = 7.2 \times 10^{-4}$
$K_4 =$	4.68×10^{-11}	$H4 = 6.1 \times 10^{-5}$
$K_5 =$	1.77×10^{-5}	$H5 = 2.2 \times 10^{-3}$
$K_6 =$	2.75×10^3	
$K_7 =$	1.3×10^6	
$K_w =$	1.0×10^{-4}	

where

$$[H^+]_T = [H_0^+] + [H_{diss}^+] \tag{15}$$

$$[OH^-]_T = [OH_0^-] + [OH_{diss}^-] \tag{16}$$

and the subscript "0" refers to prewashout concentrations and the subscript "diss" to the concentrations due to the dissociation of absorbed gases.

Gas Absorption Model

The gas absorption process can be described in terms of the physical mass transfer coefficients when the driving forces are expressed in terms of the total liquid-phase concentrations (i.e., both dissociated and undissociated forms) [7]. If the diffusivities of the various ions are assumed equal, then the mass transfer process using film theory and neglecting multicomponent diffusion effects can be expressed as

$$J_j = k_{gj} (C_{gj_b} - C_{gj_i}) \tag{17}$$

$$= k_{\ell j}^O (\phi_{j_i} - \phi_{j_b}), j = 1,2,3,4,5 \tag{18}$$

where

$$
\begin{aligned}
J &= \text{the flux} \\
j &= \text{the species (i.e., } j = 1 \text{ refers to } SO_2, \text{ etc.)} \\
C &= \text{gas phase concentrations} \\
k_g \text{ and } k_\ell &= \text{gas phase and liquid phase mass transfer coefficients} \\
i \text{ and } b &= \text{interfacial and bulk values, respectively} \\
\phi &= \text{total liquid compositions, i.e.,}
\end{aligned}
$$

$$\phi_1 = [SO_2 \cdot H_2O] + [HSO_3^-] + [SO_3^{2-}] \tag{19}$$

$$\phi_2 = [CO_2 \cdot H_2O] + [HCO_3^-] + [CO_3^{2-}] \tag{20}$$

$$\phi_3 = [NH_3 \cdot H_2O] + [NH_4^+] \tag{21}$$

$$\phi_4 = [HNO_3 \cdot H_2O] + [NO_3^-] \tag{22}$$

$$\phi_5 = [HCl \cdot H_2O] + [Cl^-] \tag{23}$$

The gas-phase mass transfer coefficients can be estimated from the Froessling equation

$$N_{Sh_j} = 2k_{gj}r/D_{gj} = 2 + 0.552 \, N_{Re_j}^{1/2} \, N_{Sc_j}^{1/3} \tag{24}$$

where

N_{Re} and N_{Sc} = the Reynolds and Schmidt numbers, respectively
D = the gas-phase diffusivity

The liquid-phase mass transfer coefficient is given by

$$k_{\ell j}^O = D_{\ell j}/\delta \qquad (25)$$

where

D_ℓ = the liquid-phase diffusivity
δ = the film thickness

The above nonlinear equations can be solved simultaneously to obtain the interfacial pH, the interfacial concentrations and the flux. The bulk liquid concentrations at a new time, $\phi_{j b}^{m+1}$, can be determined by mass balance, i.e.,

$$\phi_{j b}^{m+1} = J_j\, 3\, \Delta t/r + \phi_{j_b}^m \qquad (26)$$

where Δt is the time step and $3/r$ is the surface-area-to-volume ratio of the drop.

Results and Discussions

The described model was used to simulate the following systems:

SO_2-H_2O
HNO_3-H_2O
SO_2-HNO_3-H_2O
SO_2-NH_3-H_2O
SO_2-NH_3-CO_2-H_2O
SO_2-HNO_3-NH_3-H_2O
SO_2-NH_3-CO_2-HNO_3-HCl.

At present, these simulations are being conducted to identify the major features of precipitation scavenging from a mixture of gases. It is important to understand gas absorption by precipitation because gas absorption can occur in-cloud and below-cloud and is a necessary step before aqueous oxidation can occur. Simple "ideal" situations are studied so that the major features can be isolated and identified. In this section a sample of the results obtained is presented (a complete description of the results is presented in [8]).

In general, the results indicate that the physical parameters such as drop size, gas phase diffusivity, ambient concentration and solubility are most important in determining the relative effectiveness of the gas absorption. The effect of initial pH seems to be significant only for systems which lack NH_3. For systems lacking NH_3, the higher the initial pH the higher the absorption rates.

The presence of NH_3 seems to overshadow the initial pH effect. NH_3 increases the equilibrium pH of the drop (see Figure 1) and dramatically increases the droplet's capacity to gas scavenge. The effect of HNO_3 tends to be in the opposite direction, reducing the increase in pH and the total S(IV) concentrations caused by the presence of ammonia. This is shown in Figure 2.

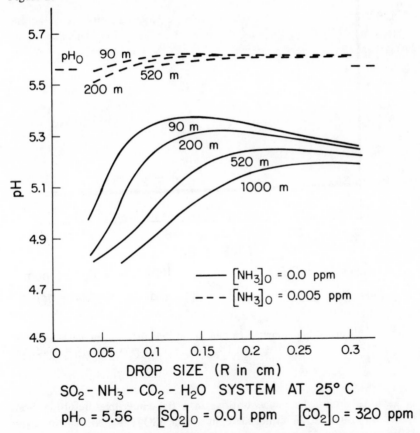

Figure 1. Droplet pH as a function of drop size for different fall distances for a SO_2–NH_3 – CO_2–H_2O system at 25°C with $[SO_2]_0$ = 0.01 ppm, $[CO_2]_0$ = 320 ppm and pH_0 = 5.56.

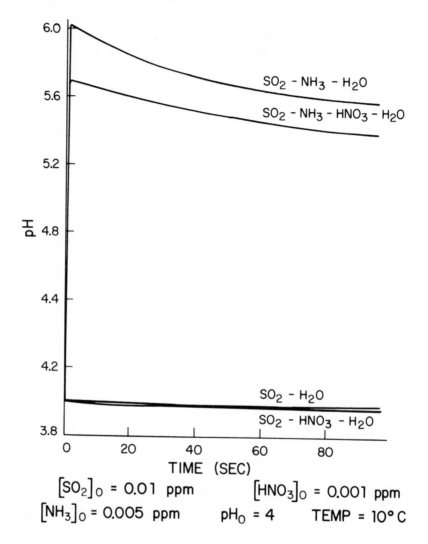

Figure 2. pH of a 0.1-cm droplet as a function of time for different ambient systems at $10°C$ and $[SO_2]_0 = 0.01$ ppm, $[HNO_3]_0 = 0.001$ ppm, $[NH_3]_0 = 0.005$ ppm and $pH_0 = 4$.

Table 2 shows the concentrations in a 0.2-cm droplet resulting from 60-sec exposure to an atmosphere in which $[SO_2] = [NH_3] = [HC] = [HNO_3] = 1$ ppb. The high gas phase diffusivity is the principal reason for the rapid absorption of NH_3. (From Equation 24 it can be seen that k_g increases with D_g.) Solubility is also an important factor. Although the gas-

Table 2. Total Liquid Concentrations in a 0.2-cm Droplet with pH_0 = 7 at a Temperature of 25°C After Exposure for 60 sec to an Atmosphere with SO_2 = NH_3 = HCl = HNO_3 = 1 ppb and CO_2 = 320 ppm

	μmol/L	$D_g cm^2$/sec	mol/L gas	
			mol/L	solution
ϕHN_3	1.5	0.234	7.2 x 10^{-4}	
ϕHCl	1.1	0.189	2.2 x 10^{-3}	
ϕHNO_3	0.8	0.132	6.1 x 10^{-5}	
ϕSO_2	0.75	0.136	3.3 x 10^{-2}	

phase diffusivities of HNO_3 and SO_2 are similar, the high Henry's law constant of HNO_3 results in HNO_3 being absorbed in higher quantities. The high Henry's law constant combined with the high dissociation constant of HNO_3 makes HNO_3 a major contributor to the acidity of precipitation (provided there is sufficient HNO_3 in the air mass). This shown in Figure 3, where under the same conditions HNO_3 lowers the pH of raindrops more than SO_2.

These results are consistent with the ambient ground level measurements reported by Georgii [9] and presented in Table 3. NH_3 and HCl are removed to a significant extent, with NH_3 showing a higher percentage decrease. The high ambient SO_2 concentrations (compared to HCl) overshadow the diffusivity effects and make SO_2 the major contributor to precipitation acidity.

Droplet pH is also strongly dependent on drop size as illustrated in Figures 1, 3 and 4. For low NH_3 concentrations the smaller drops and the larger fall distances have lower pH values. The smaller droplets have the largest surface area-to-volume ratio and therefore absorb at a faster rate. The larger fall distances imply longer exposure times, thus these droplets are closer to equilibrium. The maximum shown in the pH versus drop size curves at a given fall distance is due to the dependency of parameters such as k_g and $k\varrho$ and the terminal velocity on drop size.

The effects of NH_3 on droplet pH are shown in Figures 1, 5 and 6. The presence of NH_3 increases the solution pH at all temperatures and drop sizes. As shown in Figures 5 and 6, above a certain NH_3 level, the pH of the droplet initially increases and at still higher concentrations, continues to increase. In all cases, the lower the temperature the higher the absorption. With relatively high NH_3 concentrations, intermediate size droplets have the lowest pH values, as shown in Figure 4. This behavior is opposite of that in the absence of NH_3.

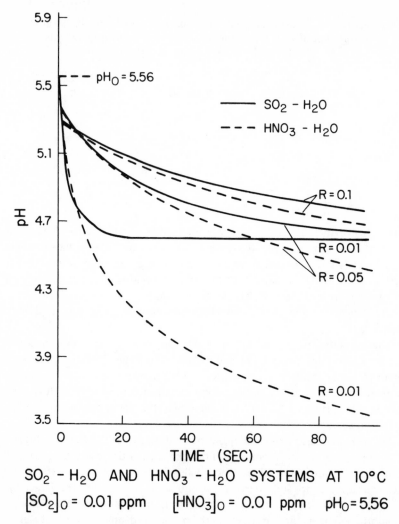

Figure 3. Droplet pH as a function of drop size for different fall distances for SO_2 – H_2O and HNO_3 – H_2O systems at $10°C$ with $[SO_2]_0 = 0.01$ ppm, $[HNO_3]_0 = 0.01$ ppm, and $pH_0 = 5.56$.

Table 3. Gas Concentrations in Ground Air.

Air Concentrations ($\mu g/m^3$)	NH_3	SO_2	Cl
Before rain	21.6	328	14.3
After rain	11.0	212	5.3
Percent remaining	52	64	39

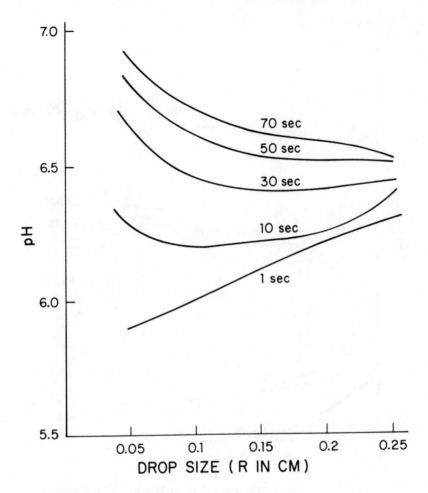

Figure 4. Droplet pH as a function of drop size for different travel times for a SO_2 – NH_3 – CO_2 – HNO_3 – HCl – H_2O system at $25^{\circ}C$ with $[SO_2]_0 = 0.01$ ppm, $[NH_3]_0 = 0.02$ ppm, $[HNO_3]_0 = [HCl]_0 = 1$ ppb, $[CO_2]_0 = 320$ ppm and $pH_0 = 5.56$.

NONISOTHERMAL EFFECTS ON SO_2 ABSORPTION

In the first part of this chapter, trace gas absorption by raindrops was discussed. This analysis concentrated on the simultaneous absorption of many trace gases and how the presence of other trace gases affected the absorption processes. This analysis isolated these effects by treating the raindrop as an isothermal-constant-sized sphere. However, in the real environment a raindrop behaves dynamically as it falls through the atmosphere and the temperature

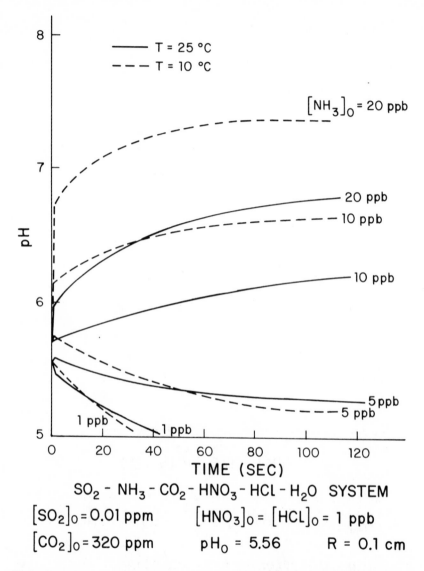

Figure 5. pH of a 0.1-cm droplet as a function of exposure time for a SO_2 – NH_3 – CO_2 – HNO_3 – HCl – H_2O system with $[SO_2]$ = 0.01 ppm, $[HNO_3]_0$ = $[HCl]_0$ = 1 ppb, $[CO_2]_0$ = 320 ppm and pH_0 = 5.56.

of the drop as well as the radius changes during descent. So it is also important to characterize the influence of these nonisothermal effects on the absorption process. These effects were studied by modeling SO_2 absorption by a drop

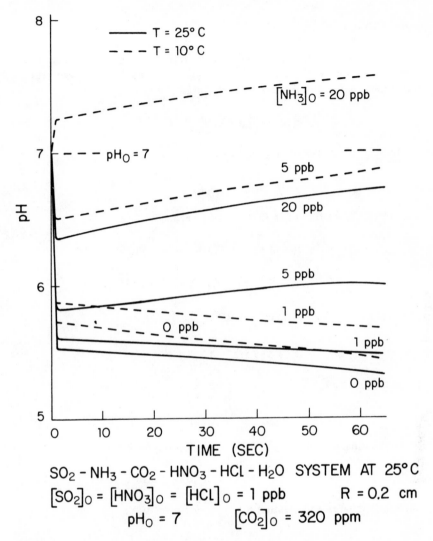

$SO_2 - NH_3 - CO_2 - HNO_3 - HCl - H_2O$ SYSTEM AT 25° C

$[SO_2]_0 = [HNO_3]_0 = [HCl]_0 = 1$ ppb   R = 0.2 cm

$pH_0 = 7$ $[CO_2]_0 = 320$ ppm

Figure 6. pH of a 0.2 cm droplet as a function of exposure time for a $SO_2 - NH_3$ - $CO_2 - HNO_3 - HCl - H_2O$ system with $[SO_2]_0 = [HNO_3]_0 = [HCl]_0 = 1$ ppb, $[CO_2]_0 = 320$ ppm and $pH_0 = 7$ or 4.

undergoing evaporation. The analysis includes the effects of droplet evaporation and the change in physical properties and reaction rates with temperature, and treats multicomponent diffusion.

Model Description

The dynamic equations of interest for a droplet as it travels between cloud base and ground level are those that describe the bulk S(IV) concentration, the droplet radius, and the droplet temperature, i.e.,

$$d\phi_b/dt = 3/r \ J^g_{SO_2} \tag{27}$$

where $J^g_{SO_2}$ is the flux of SO_2 to the surface,

$$dr/dt = J^g_{H_2O}/\rho_L \tag{28}$$

where $J^g_{H_2O}$ is the flux of H_2O to the surface, and

$$dT_{drop}/dt = [3/(4.18 \times 10^7)] \ 1/r \ [a/(e^a - 1) \ k/r \ (T_B - T_{drop} \ e^a)$$

$$+ \alpha J^g_{H_2O} \lambda + 10^{-3} \ J^g_{H_2O} \ (C^g_{P_w} - C^L_{P_w}) \ (T_B - T_{drop})$$

$$+ 10^{-3} \ J^g_{SO_2} \ H^g_{SO_2} + 10^{-3} \ J^g_{SO_2} \ C^g_{P_s} \ (T_B - T_{drop})$$

$$+ 0.95 \ \sigma \ (T_B^4 - T_{drop}^4)] \tag{29}$$

where a is given by

$$a = r/k \ [J^g_{SO_2} \ C^g_{P_s} + J^g_{H_2O} \ C^g_{P_w}] \cdot 10^{-3} \tag{30}$$

k is the thermal conductivity in cal/cm-sec-°K given by

$$k = 5.49 \times 10^{-5} + 1.68 \times 10^{-7} \ (T - 273.16) \tag{31}$$

and λ is the heat of vaporization for the water given by

$$\lambda = 753.68 - 0.5725T \tag{32}$$

These equations are coupled and must be solved simultaneously.

The flux of SO_2 to the droplet is determined by assuming that at any time the flux of SO_2 from the gas phase, $J^g_{SO_2}$, must equal the flux of absorbed SO_2 (and dissociated forms) in the liquid phase, N^*_ϕ, i.e., (where "*" denotes interfacial values).

$$J^g_{SO_2} = N^*_\phi \tag{33}$$

A similar equation is assumed to hold for H_2O.

$$N^*_{H_2O} = J^g_{H_2O} \tag{34}$$

When there are high H_2O mass transfer rates, as those that occur during droplet evaporation or condensation, the flux of H_2O can have a significant influence on the flux of SO_2. This effect is modeled by use of the Stefan-Maxwell equations which describe ordinary diffusion in multicomponent gases. For the SO_2-H_2O-N_2 system these equations can be solved to yield

$$J^g_{SO_2} = N_{sh}/2\,\bar{J}\left[\frac{P^B_{SO_2}\exp-(\bar{J}\,r/C_f\,D^g_{SO_2})-P^*_{SO_2}}{\exp-(\bar{J}\,r/C_fD^g_{SO_2})-1}\right] \tag{35}$$

$$J^g_{H_2O} = N_{sh}/2\,\bar{J}\left[\frac{P^B_{H_2O}\exp-(\bar{J}\,r/C_f\,D^g_{H_2O})-P^*_{H_2O}}{\exp-(\bar{J}\,r/C_f\,D^g_{H_2O})-1}\right] \tag{36}$$

and

$$\bar{J} = -C_f\,D^g_{N_2}/r\,\ln(y^*_{N_2}/y^B_{N_2}) \tag{37}$$

where C_f is the total gas concentration.

The expression for the total flux of S(IV) taking into account multi-component diffusion in the liquid is

$$N^*_\phi = (\phi^*/CT)N^*_{H_2O} + k^0_L(\phi^* - \phi^B) - k^0_L\,A([HSO_3^-]^* - [HSO_3^-]^B) \tag{38}$$

where CT is the total liquid concentration and

$$A = 1 - \frac{D^L_{HSO_3^-}}{D^L_{SO_2}} \tag{39}$$

(where here it has been assumed that HSO_3^- is the only dissociated form of SO_2 in solution).

The electrical-neutrality condition can be used to solve for the interfacial compositions of H^+, HSO_3^- and $SO_2 \cdot H_2O$, i.e.,

$$[H^+] = [HSO_3^-] + C_{Ex} + K_w/[H^+] \tag{40}$$

and

$$[HSO_3^-] = [-\psi + \sqrt{\psi^2 + \delta K1 \phi}]/2 \tag{41}$$

where C_{Ex} is the concentration of the initial strong acid or base and is positive for a strong acid, negative for a strong base and zero at pH 7, and

$$\psi = 2K1 + C_{Ex} + \sqrt{([HSO_3^-] + C_{Ex})^2 + 4K_w} \tag{42}$$

and

$$[SO_2 \cdot H_2O] = \phi - [HSO_3^-]. \tag{43}$$

These equations comprise the basic model for the nonisothermal absorption of SO_2. These equations can be solved simultaneously to yield the droplet radius, temperature, pH, and bulk and interfacial S(IV) concentration profiles as a function of fall distance.

Simulations

This model has been used to simulate SO_2 absorption below-cloud by raindrops falling through a 2-km-high layer in the atmosphere near the surface of the earth in which sulfur dioxide is distributed in some specified manner. The temperature profile is assumed to be adiabatic with a ground temperature of 23°C and the ambient relative humidity is assumed to be 80%. The raindrops are assumed to be spherical and to fall vertically at their terminal velocities. Before entering this layer, the droplets attain an initial pH designated as pH_0.

Equations 27–30 were solved using the Gear method and Equations 33–43 were solved using the Newton-Raphson method. The physical properties of the system were reevaluated as the droplet temperature changed.

Results and Discussions

A summary of the simulation results is presented in Table 4. It can be seen from these results that droplets smaller than ~0.08 cm reach the surface with a temperature of about 20°C. This temperature is below the ambient ground level value of 23°C. This illustrates that there is sufficient evaporation occurring to maintain these droplets at a temperature below ambient. (The droplets do, however, follow the ambient profile.) The temperature of larger droplets is lower due to their larger mass and their larger terminal velocities. These droplets have insufficient time to respond completely to the atmospheric profile.

Evidence of significant evaporation rates is present in the predicted ground-level radius values. The smaller droplets experience radial variations during descent, with the smallest droplets having the largest percentage change. Droplets as small as 0.01 cm evaporate by 90% in a fall distance of only $\sim$ 200 m, and most likely never reach the surface.

Droplet evaporation can help to explain why the smaller drops (e.g., $r \leqslant 0.06$ cm) have lower ground level concentrations for $pH_0 = 4$ than the larger drops. The concentration profiles of interfacial and bulk S(IV) for a 0.06 cm droplet are presented in Figure 7. These profiles pass through a maximum at about 800 m then decrease steadily. This decrease is due in part to the evaporation and in part to the fact that the droplet is increasing in temperature (thus the solubility of the gas is decreasing).

This desorption phenomena has been demonstrated experimentally by Flack and Matteson [10]. Their measured bulk S(IV) concentration versus time curves for various supersaturation ratios are presented in Figure 8. Their results also showed that the droplet temperature was below the ambient value.

It can be seen from Table 4 that the higher the initial pH the higher the ground level S(IV) concentrations. For large drops absorption is quite slow and a saturation level is never reached during the 2-km fall distance (see Figure 9). For these larger drops, the pH change is slight. However, for smaller drops ($r \leqslant 0.04$ cm) sufficient SO_2 is absorbed to reach a break-through point and the pH of the droplet changes dramatically. The pH and S(IV) profiles as a function of fall distance for a 0.01-cm droplet are shown in Figures 10 and 11. In this case breakthrough occurs at $\sim$ 40 m with the pH falling to 5.3. The S(IV) concentration remains constant after breakthrough until rapid desorption occurs beyond $\sim$ 150 m.

Simulations have also been conducted for atmospheres in which the ambient SO_2 profiles are nonuniform. one case studied was that where the ambient SO_2 value is 1 ppm from cloud base to 1600 m, then changes to 10 ppb from 1600 m to ground level. This is similar to a raindrop falling through an elevated plume. Simulation results sowed that for $pH_0 = 10$ the concentration changed, from 240 μmol/L at 1600 m to 170 μmol/L at ground level. This desorption leads to a modest increase in pH (see Figure 12), but more importantly to a redistribution of ambient SO_2, with SO_2 being in effect transported from higher to lower elevations.

SUMMARY

Results obtained from the described gas absorption models demonstrate that the absorption behavior of raindrops is strongly dependent on drop

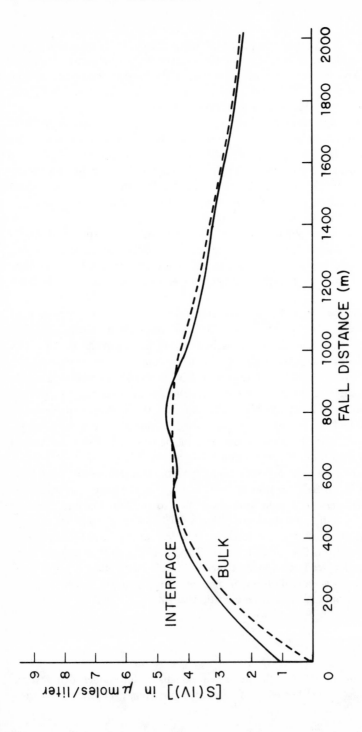

Figure 7. Interfacial and bulk S(IV) concentrations for a 0.06 cm (initial radius) droplet with $pH_0 = 4$ falling through an adiabatic atmosphere. Ground level corresponds to 2000 m.

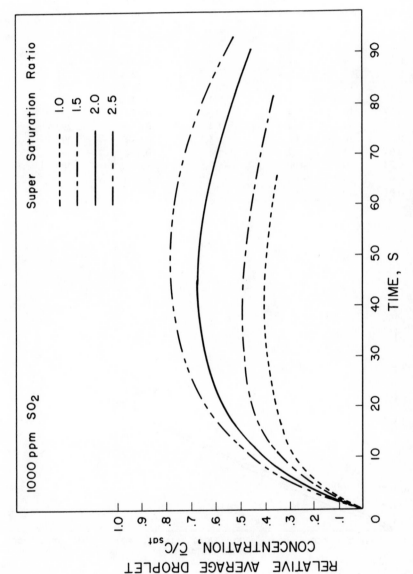

Figure 8. Total sulfur concentrations measured by Flack and Matteson [10] for a 0.23-cm droplet exposed to a humid nitrogen stream with 1000 ppm SO₂.

Table 4. Results for Nonisothermal Simulations

Initial Drop Size (cm)	pH$_0$	Ground Level S(IV) (μmol/L)	Ground Level pH	Ground Level Radius (cm)	Ground Level Temperature (°K)
0.5	4	0.25	~4	~0.5	281
0.5	10	0.35	~10	~0.5	281
0.1	4	2.5	~4	0.0998	291
0.1	7	5.4	5.2	0.0998	291
0.1	10	6.3	~10	0.0998	291
0.08	4	3.0	~4	0.079	293
0.08	10	24	~10	0.079	293
0.06	4	2.0	~4	0.058	292.5
0.06	10	45	4.7	0.058	292.5
0.04	4	2	~4	0.034	293
0.04	10	100	3.5	0.034	293
0.01	4	—	—	~0	—
0.01	10	—	—	~0	—

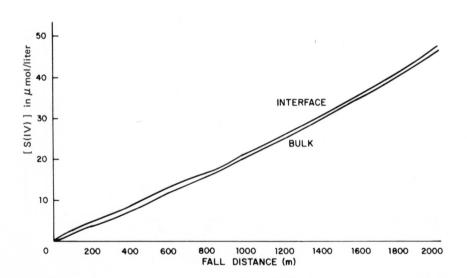

Figure 9. Interfacial and bulk S(IV) concentrations for a 0.06-cm (initial radius) droplet with pH$_0$ = 10.

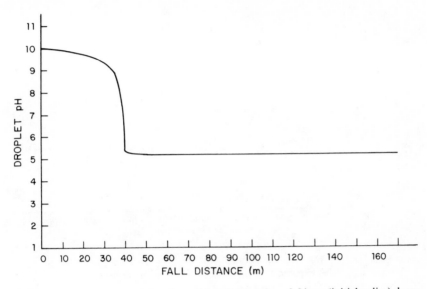

Figure 10. Droplet pH as a function of fall distance for a 0.01-cm (initial radius) droplet and $pH_0 = 10$.

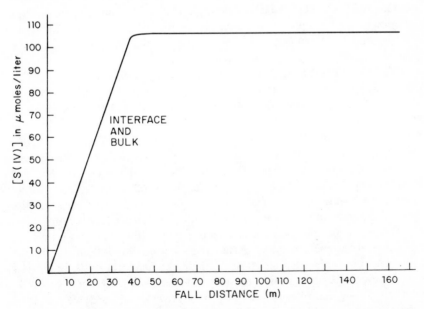

Figure 11. Interfacial and bulk S(IV) concentrations as a function of fall distance for a 0.01-cm (initial radius) droplet and $pH_0 = 10$.

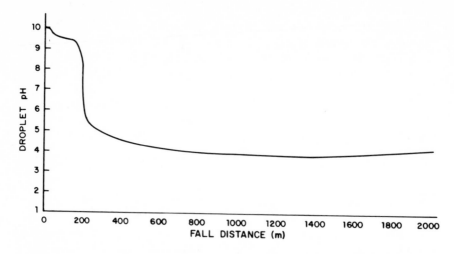

Figure 12. Droplet pH as a function of fall distance for a 0.1-cm (initial radius) droplet and $pH_0 = 10$ exposed to 1 ppm from cloud base to 1600 m and 10 ppb from 1600 m and 10 ppb to ground level.

size, droplet temperature, fall distance and ambient profiles of temperature, relative humidity and trace gas concentrations. These parameters are important for both chemical and physical processes. Thus an accurate analysis of ground level samples in terms of the contribution of gas scavenging to the measured pH requires that field studies obtain information on these parameters.

SO_2 appears to be the major contributor to precipitation acidity resulting from gas scavenging by virtue of its high ambient concentrations. HNO_3 is highly soluble and dissociates rapidly in solution and can be a significant contributor to solution acidity. The presence of NH_3 in the atmosphere increases solution pH and the rain's capacity to gas scavenge. The effect of NH_3 is highly dependent on the ambient concentration and its influence can change during the rain event. In the initial stages of the rain the ambient concentrations are at their highest values; however since NH_3 is rapidly removed by water droplets (due to its high gas-phase diffusivity and solubility), the ambient values can fall significantly in a relatively short time.

For the same set of initial and ambient conditions, samples located at different elevations can record different chemical compositions due to the differences in droplet fall distances. From an isothermal analysis the samples at the lower elevation should record the lower pH (for low ambient NH_3 concentrations) regardless of the drop sizes. However, if the dynamics of the drop are considered, and if the rain was initially acidic and comprised mostly of small droplets, then the samples at the higher elevation would record the lower pH. This is due to nonisothermal effects and the fact that small drops, although they absorb quickly, have significant evaporation during descent. In addition, this can affect their volume contribution to a bulk sample.

Results also indicate that drops respond quickly to changes in ambient values and thus droplets passing through elevated plumes can absorb significant amounts in-plume and then desorb below-plume. This can lead to modest increases in ground level pH, but more importantly to a redistribution of ambient concentrations, with gases being, in effect, transported from higher to lower elevations.

ACKNOWLEDGMENT

This research was supported in part by the National Aeronautics and Space Administration under Research Grant NAG1-36.

REFERENCES

1. Likens, G. E., F. W. Richard and J. N. Galloway. "Acid Rain," *Scientific Am.* 241:43–51 (1979).
2. Terraglio, F. P., and R. M. Manganelli. "The Absorption of Atmospheric Sulfur Dioxide by Water Solutions," *J. Air Poll. Control. Assoc.* 17: 403–406 (1967).
3. Hill, F. B., and R. F. Adamowicz. "A Model for Rain Composition and the Washout of Sulfur Dioxide," *Atmos. Environ.* 11:917–927 (1977).
4. Adamowicz, R. F. "A Model for the Reversible Washout of Sulfur Dioxide, Ammonia and Carbon Dioxide from a Polluted Atmosphere and the Production of Sulfates in Raindrops," *Atmos. Environ.* 13: 105–121 (1979).
5. Overton, J. H., V. P. Aneja and J. L. Durham. "Production of Sulfate in Rain and Raindrops in Polluted Atmospheres," *Atmos. Environ.* 13:355–367 (1979).
6. Aneja, V. P., H. J. Overton and J. L. Durham. "SO_2 Flux to a Falling Raindrop in a Polluted Atmosphere," *Am. Inst. Chem. Eng. J.* 75: 151–155 (1979).
7. Carmichael, G. R., and L. K. Peters. "Some Aspects of SO_2 Absorption by Water-Generalized Treatment," *Atmos. Environ.* 13:1505–1513 (1979).
8. Adewuyi, Y. G. "The Influence of Gas Scavenging of Trace Gases on Precipitation Acidity," MS Thesis, Chemical Engineering Program, University of Iowa, Iowa City, IA (1980).
9. Georgii, H. W. "Untersuchungen über atmosphärische Spurenstoffe und ihre Bedeutung für die Chemie der Niederchlage," *Geofis. pura appl.* 47: 155–171 (1960).
10. Flack, W. W., and M. J. Matteson. "Mass Transfer of Gases to Growing Water Droplets," *Polluted Rain,* T. Y. Toribara, Ed., (New York: Plenum Publishing Corporation, 1980), pp. 61–83.

CHAPTER 16

PRELIMINARY RESULTS OF
AN EIGHT-LAYER REGIONAL ASSESSMENT MODEL
APPLIED TO THE PROBLEM OF ACID RAIN

W. E. Davis

Battelle Pacific Northwest Laboratory
Richland, Washington 99352

INTRODUCTION

One of the problems confronting the northeastern United States is acid rain. Since acid rain is produced in convective storms as well as in frontal storms, it is necessary to understand how the atmospheric transport of SO_2 and sulfates from the sources interacts with the convective and frontal storms to produce the acid rain. At present most attempts at modeling the transport of SO_2 and sulfates have used single-layer models [1, 2, 3, 4]. The single layer models use single-layer averaged u and v wind components to transport the SO_2 and sulfates. Such an approach can lead to erroneous conclusions, especially when used near a frontal storm (Figures 1 and 2), due to the effect of the wind shear that occurs in these storms [5].

An eight-layer diabatic model has been developed to look at the problem of transporting air parcels through frontal storms [6]. However this model has not been tested to see how it performs in the modeling of SO_2 and sulfate air concentrations and depositions. Before running these tests for cases using SO_2 and sulfates, a decision was made to compare the models' trajectories

287

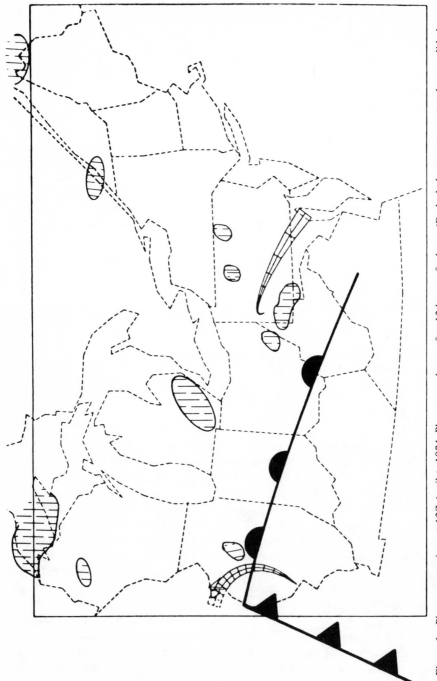

Figure 1. Plume comparison at 12Z April 1, 1974. Plumes are shown after 12 hours of release. (Dashed plumes are constant layer, solid plumes are isentropic and stippled areas indicate observed precipitation.)

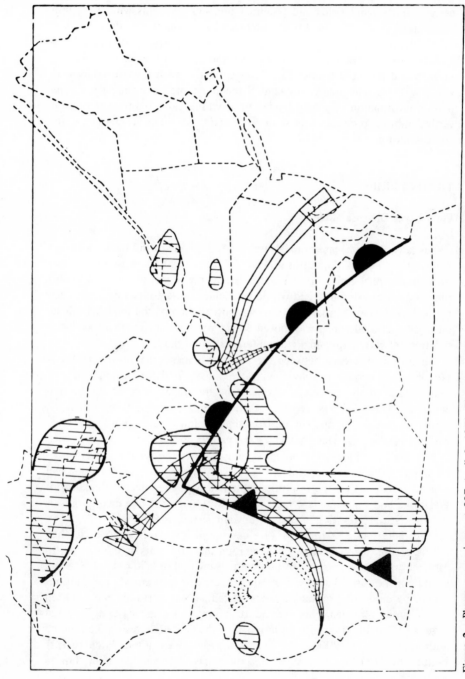

Figure 2. Plume comparisons at 00Z April 2, 1974. Plumes are shown after 24 hours of release. (Dashed plumes are constant layer, solid plumes are isentropic and saturated air parcels are denoted by *.)

using u and v wind fields and potential temperature fields with those produced using u, v and w wind fields produced by a predictive model developed by Kreitzburg and Rasmussen [7]. The primary purpose of such a comparison would be to see how three-dimensional trajectories computed using the u, v and potential temperature fields compared with trajectories using u, v and w fields generated by dynamical predictive model. From such a comparison information may be gained as to why using potential temperature for vertical motion produces similar or dissimilar trajectories from using w for vertical motion.

TRAJECTORIES

Eight-Layer Diabatic Model

Basically the eight-layer model uses u and v gridded wind fields, potential temperatures and mixing ratios gridded fields for eight layers in order to run three-dimensional trajectories. For this study the model was altered to also run trajectories using u, v and w gridded fields. These gridded fields for each of the layers were then used in the transport section of the model. Each sequentially released parcel was moved with a wind obtained with a time interpolation between maps and a space interpolation in the grid [8].

The major difference between the two techniques is that horizontal transport is calculated in the layer with the most closely matching potential temperature to the parcels' potential temperature at each advection step along the trajectory. The second method integrates the vertical motion, w, along the parcel path and the horizontal advection is calculated in the layer with the height most nearly matching the parcels.

For the technique using potential temperatures the vertical movement is by stepping the parcel up or down. The accuracy of this approximation will depend somewhat on the resolution obtained with the selected thickness of the layers and on diabatic effects along the trajectory. The mixing ratio is used to test for saturation in cases of upward vertical motion.

The conditions of saturation were monitored for two reasons. First, they are used, along with observed precipitation measurements at the surface, to estimate whether scavenging of pollutants is incloud or below cloud. Second, saturation indicates the release of latent heat, which means vertical motions should be based on equivalent potential temperatures. Hourly precipitation values can be obtained from over 3000 sites in the United States as shown in Figure 3. For this study, the eastern half of the United States was considered with approximately 2000 sites. The hourly data were interpolated to $1/2°$ latitude by $1/2°$ longitude sampling grid boxes. An example of the hourly values is shown in Figure 4.

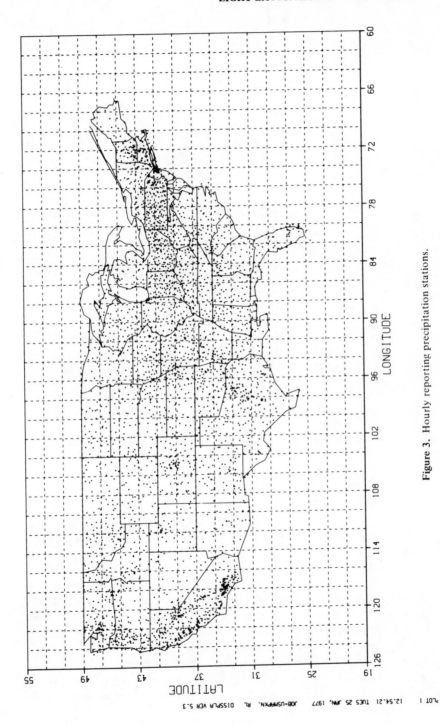

Figure 3. Hourly reporting precipitation stations.

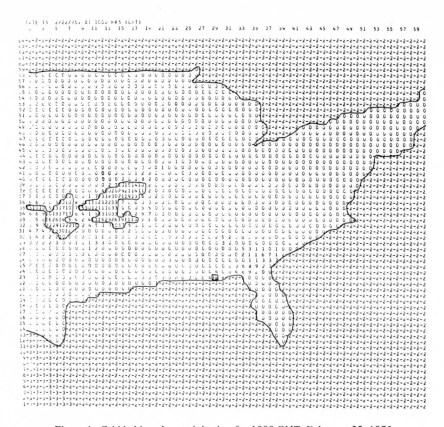

Figure 4. Gridded hourly precipitation for 1000 GMT, February 22, 1975.

Beyond calculating saturation in the model, a number of significant diabatic effects have been estimated within the framework of the model. The most predominant of these are the climatological effects of radiational heating and cooling, as well as mixing in the lower layers. The magnitude of these diabatic effects will vary with the season of the year. Others, such as cloud cover, suspended particulates and mixing, aerosols, real-time changes in the temperature due to radiational heating and cooling convection, and wind shear over the vertical extent of the parcel, have not been included. These effects cause concern because they can introduce varying degrees of error in the vertical motion determinations. Work is continuing to establish techniques to account for the diabatic effects which are determined to be affecting model results significantly. Initially, in an attempt to reduce the impact of the near-surface diabatic effects, the lower layer began 100 m above the surface. The climatological changes in diurnal temperature with height were estimated from tower temperature data as well as some mountain temperature data. These data were used to add a diurnal cycle to the potential temperature of the released puff.

Data Input

Kreitzberg Model

The Kreitzberg model has been described in Kreitzberg and Rasmussen [7] and in Perkey and Kreitzberg [9]. The model is a limited area mesoscale primitive equation model which produces u, v and w wind components, rain water, cloud water, precipitation, pressure and virtual temperature, as well as additional cloud information. The data fields were extracted from a 36-hr forecast with fields produced every three hours for the test case using 8 of the 16 layers in the model. Potential temperature fields were calculated from the above data for the same eight layers.

Test Case

A severe outbreak of tornados occurred on April 4, 1974 on the eastern seaboard. Associated with this outbreak was a frontal storm shown on surface maps for April 3, 12Z and April 4, 12Z (Figures 5 and 6).

Kretizberg's model was run for a 36-hr period in this storm, from April 3, 00Z to April 4, 12Z, 1974. Gridded fields for every three hours were obtained for a 1.25 latitude and a 1.56 longitude grid. These data were then used in the framework of the eight-layer diabatic model. Data at the levels 25, 375, 750, 1250, 2000, 3000, 4500 and 6000 meters above surface were used.

Comparisons of trajectories for a series of releases at 12 different source points shown in Figure 7 were made for a 36-hr period using two different options. The first set was constructed using Krietzberg's three-hour data fields of u, v and w and potential temperature to compute trajectories. The second set used Kreitzberg's 12-hr data fields instead of 3-hr fields in the diabatic model.

RESULTS

Trajectories were computed on the eight-layer model for the following data: (1) 3-hr fields of u, v and w, (2) 3-hr fields of u, v and potential temperature, (3) 12-hr fields of u, v and potential temperature and (4) 12-hr fields of u and v (constant level). The last three were compared with the first one.

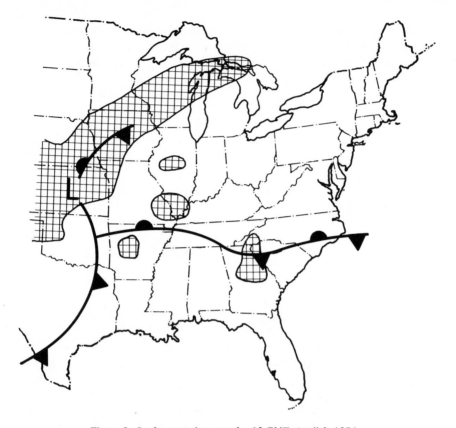

Figure 5. Surface weather map for 12 GMT, April 3, 1974.

The trajectories' starting points are shown on Figure 7 and the time of study was for the 36-hr period of April 3, 00Z to April 3, 12Z, 1974.

The results from running trajectory from the release points shown in Figure 7 are shown in Figure 8, which plots the mean distance separation against the number of hours after release. The reader should note that for the one hour after release, 36 data points were available, but at 36 hours after release only one data point was available. Because of the lack of data after 24 hours, only ≤ 24 hours after release were considered. This means by using the whole 36 hours that 12 data points could be compared for 24 hours after release.

In general the results show that the mean end point differences vary as a function of distance from the storm center, as can be seen in Figure 8, when compared to Figures 5 and 6. The greater the distance from where the storm center passed, the less the end point differences. The results showed that up

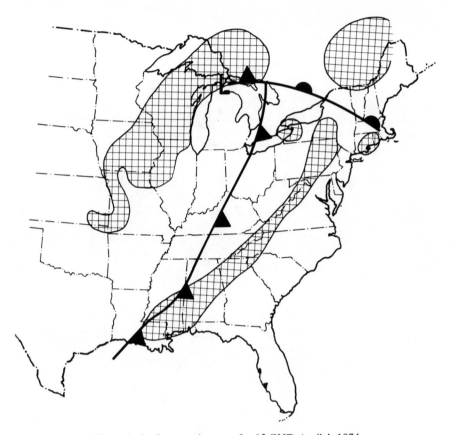

Figure 6. Surface weather map for 12 GMT, April 4, 1974.

to 24 hours the end points generated using u, v and potential temperature compared more favorably with the end points generated using u, v and w than the end points generated using only u and v (constant level).

The reader should note that w, vertical motion, is the mean air mass vertical motion. Since mixing is not incorporated along the path of the parcel, large differences in potential temperature between the parcel and the environment can occur. When using these differences to calculate vertical motion, an overestimate in the height rise of a parcel can occur. This means that those parcels rising to higher levels will be transported at generally higher wind speeds at that level causing large difference in end points. Also it was found that significant differences in trajectory end points occurred between using the forecast u, v and w fields and the fields of u, v and potential temperature generated using radiosonde data (Figure 8b). This was

Figure 7. Trajectory release points.

especially true for source points near the center of the low. A variation of only a few degrees near the low resulted in trajectory end points with large differences. Another result, which is quite important, is shown in Table 1. More of the eight-layer models' trajectories using u, v and potential temperature approach showed more vertical movement than the trajectories using the u, v, w fields. This was attributed to the w fields which are indicative of the mean movement of that layer rather than a point movement. Further, the isentropic approach only considers a parcel as a point and by not allowing mixing along the trajectory causes the trajectory to be overly sensitive to vertical movement. Also when saturation occurs, the parcel is moved vertically based on the equivalent potential temperature. As mentioned be-

Table 1. Comparison of Trajectory Levels

Hours After 4/3/00Z	Difference From Release (%)			
	K[a]	8[b]	K + 8[c]	>| 1 Level |[d]
A. Level 2 Releases (~375 meters above surface)				
6	10	25	20	0
12	10	37	37	1
18	20	63	56	9
24	40	50	68	8
B. Level 4 Releases (~1250 meters above surface)				
6	10	20	13	0
12	11	20	14	0
18	17	40	32	0
24	34	40	45	8

[a]Using Kreitzberg's data u, v, w.
[b]Using Kreitzberg's data u, v and calculated potential temperature.
[c]Percent different from levels calculated using u, v, w data.
[d]Percent greater than 1 level different from levels calculated using u, v, w.

fore, in a considerable number of cases this occurred moving the parcel to the top layer in the model. In reality this may or may not happen due to the amount of mixing of the air parcel with the environmental air. This mixing will alter the potential temperature of the parcel and cause reduced vertical motions in a number of cases. At the present time research is under way in attempting to better estimate mixing along the path of the eight-layer models' trajectories using potential temperature.

CONCLUSIONS

Trajectories using the u, v and potential temperature compared more favorably in some cases with trajectories computed using fields of u, v and w than did those using constant layer winds over a 36-hr period. The trajectories produced using the u, v and potential temperature were shown to move vertically more rapidly than trajectories using the u, v, w fields generated by Kreitzberg's model. This rapid vertical movement was attributed to the lack of mixing with environmental air along the trajectories' path. More work is intended to incorporate into the model mixing along the parcels' path. Work

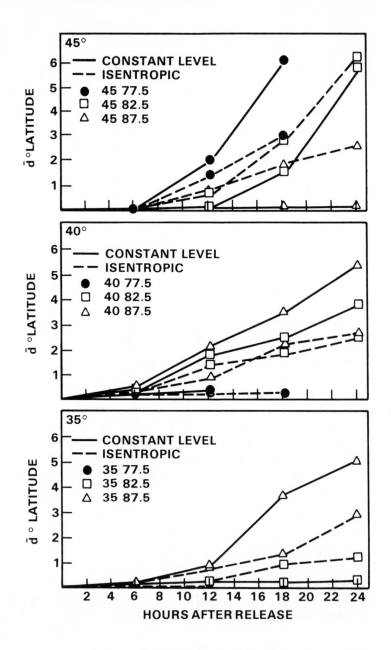

Figure 8. 3-hr data mean distance separation between trajectory end points as a function of time after release. (●, □ and △ indicate release points.)

also is now under way comparing the eight-layer models' results with observed air concentrations and wet and dry deposition for SO_2 and sulfates for October 1977.

ACKNOWLEDGMENTS

The author wishes to thank Dr. Carl Kreitzberg and Dr. Don Perkey for providing the data tapes for use in the study.

REFERENCES

1. Bolin, B., and C. Persson. "Regional Dispersion and Deposition of Atmospheric Pollutants with Particular Application to Sulfur Pollution over Western Europe," *Tellus XVII*, 3:281–310 (1975).
2. Brookhaven National Laboratories. "Annual Report, FY 1975," *Regional Energy Studies Program*, P. F. Palmedo, Ed., (1975), pp. 46–61, 91–106.
3. McNaughton, D. J. "Initial Comparison of SURE/MAP3S Sulfur Oxide Observations with Long-Term Regional Model Predictions," *Atmos. Environ.* 14(1) (1980).
4. Ottar, B. "An Assessment of the OECD Study on Long-Range Transport of Air Pollutants (LRTAP)," *Atmos. Environ.* 12(1–3) (1978).
5. Davis, W. E., and L. L. Wendell. "Some Effects of Isentropic Vertical Motion Simulation in a Regional Scale Quasi-Lagrangian Air Quality Model," preprint from the Third Symposium on Atmospheric Turbulence, Diffusion and Air Quality (1976).
6. Davis, W. E. "Comparison of the Results of an Eight Layer Regional Model Versus a Single Layer Regional Model for a Short Term Assessment," WMO Symposium on the Long-Range Transport of Pollutants and Its Relation to General Circulation Including Stratospheric/Tropospheric Exchange Processes, Sofia, Bulgaria (1979).
7. Kreitzberg, C. W., and R. G. Rasmussen. "Treatment of Cloudiness in Regional Scale Numerical Weather Prediction Models," AMS Conference on Cloud Physics and Atmospheric Electricity, Issaquah, WA, July 1978.
8. Wendell, L. L. "Mesoscale Wind Fields and Transport Estimates Determined from a Network of Wind Towers," *Mon. Weather Rev.* 100:565–578 (1972).
9. Perkey, D. J. and C. W. Kreitzberg. "A Time-Dependent Lateral Boundary Scheme for Limited-Area Primitive Equation Models," *Mon. Weather Rev.* 104:744–755 (1976).

INDEX

301